THE
REFERENCE TABLES
UNEARTHED

A Clear & Simple Reference Tables Guide

for the

New York State Earth Science Regents

Y. Finkel

Copyright © 2019 Y. Finkel

All rights reserved.

ISBN-13: 978-0-578-22373-5

This book may not be reproduced in any form without written consent from the author.

To order additional copies or for more information:

Call 718.605.4634

Call/Text 646.242.3490

Email unearthingnysesrt@gmail.com

What is This Book & How Do I Use It?

> Before you begin this examination, you must be provided with the *2011 Edition Reference Tables for Physical Setting/Earth Science*. You will need these reference tables to answer some of the questions.

Did you know that about 35-50% of every Earth Science Regents is composed of questions entirely based on the Earth Science Reference Tables?

And did you know that a raw score of approximately 50% on the Earth Science Regents converts to a scale score of 65%? (with at least $^9/_{16}$ lab credits)

If you know how to read *every* table on the Earth Science Reference Tables, that's terrific.

But what if you don't?

Gaining a clear understanding of the reference tables is crucial for the Earth Science Regents.

The good news is that one of the best-kept secrets of the Earth Science regents is that the reference tables-based questions are the *easiest part of the regents* by far – *if* you know how to use the reference tables.

That's where this book comes in. ***The Reference Tables Unearthed: A Clear & Simple Reference Tables Guide*** is a book that:
- Gives step-by-step instructions in **clear** and **simple** terms on how to easily decipher 26 charts on the Earth Science Reference Tables
- Highlights important information often asked on the Earth Science Regents
- Provides actual regents questions at the end of each section, along with answers and brief explanations

To Get the Most Out of This Book:

Read the book aloud with a friend so you don't miss anything important.

As you read through the book, follow along with a separate copy of the Earth Science Reference Tables. This way, you won't have to keep flipping pages from the tables to their explanations.

If you are pressed for time, start with the tables that appear most often on the regents. On each table, notice this icon with a number. This represents the average number of questions on that table per regents.

For example, on the Generalized Bedrock Geology of New York State map, you see . This means that there is an average of 2-3 questions per regents on this table.

After you finish reading about each table, do the **practice regents questions** on the table to ensure you understood it correctly. The practice questions are conveniently included after each section, symbolized by this icon:

In addition, the "More Practice" section at the end of the book organizes all the reference tables-based regents questions from the June 2012-June 2017 regents by table.

These extra questions will provide you with even more opportunity to exercise the Reference Tables skills you have learned from this guide. This way, you will be fully prepared to tackle those questions on your upcoming regents exam.

Good luck!

Y. Finkel

Contents

Equations ... 2
Specific Heats of Common Minerals ... 6
Properties of Water ... 7
Average Chemical Composition of Earth's Crust, Hydrosphere, and Troposphere ... 8
Generalized Landscape Regions of New York State ... 12
Generalized Bedrock Geology of New York State ... 13
Surface Ocean Currents ... 20
Tectonic Plates ... 24
Relationship of Transported Particle Size to Water Velocity ... 30
Rock Cycle in Earth's Crust ... 33
Scheme for Igneous Rock Identification ... 34
Scheme for Sedimentary Rock Identification ... 42
Scheme for Metamorphic Rock Identification ... 44
Inferred Properties of Earth's Interior ... 50
Earthquake P-Wave and S-Wave Travel Time ... 54
Dewpoint (°C) / Relative Humidity (%) ... 60
Temperature ... 66
Pressure ... 67
Key to Weather Map Symbols ... 69
   Station Model & Station Model Explanation ... 69
   Air Masses ... 73
   Tornado/Hurricane Symbols ... 73
   Fronts ... 74
Selected Properties of Earth's Atmosphere ... 80
Planetary Wind & Moisture Belts in the Troposphere ... 82
Electromagnetic Spectrum ... 86
Characteristics of Stars ... 90
Solar System Data ... 93
Properties of Common Minerals ... 100
ANSWERS ... 105
MORE PRACTICE ... 111

NEW YORK STATE
EARTH SCIENCE REFERENCE TABLES

PAGE 1

- Equations
- Specific Heats of Common Minerals
- Properties of Water
- Average Chemical Composition of Earth's Crust, Hydrosphere, and Troposphere

Equations

$$\text{eccentricity} = \frac{\text{distance between foci}}{\text{length of major axis}}$$

Equations

$$\text{Eccentricity} = \frac{\text{distance between foci}}{\text{length of major axis}}$$

$$\text{Gradient} = \frac{\text{change in field value}}{\text{distance}}$$

$$\text{Rate of change} = \frac{\text{change in value}}{\text{time}}$$

$$\text{Density} = \frac{\text{mass}}{\text{volume}}$$

- **Eccentricity** – the measure of how elliptical/eccentric (oval) a planet's orbit is. Eccentricity is expressed in **decimal** format.
    → The smaller the distance between two foci, the more *circular* the orbit. The eccentricity will be closer to 0.
    → The greater the distance between two foci, the more *eccentric* (oval) the orbit. The eccentricity will be closer to 1.

- **Foci** (plural of focus) – two points within the orbit. One is always the *sun*.

- **Major axis** – the longest line you can draw from one end of the oval to the other, passing through the two foci

- On your **lab test**, you must compare your result to the eccentricity of a given planet and state whether the planet's orbit is more eccentric or less eccentric than the orbit you were given. (Use **RT 15 – Solar System Data** – to find out what the planet's eccentricity is.)

> Throughout this book, the letters **"RT"** with a number is used as an abbreviation for that specific page on the reference tables. For example, **RT 3** stands for Reference Tables **page 3**.

- <u>Example:</u> The orbit you were given has an eccentricity of **0.345**. How does that compare to the eccentricity of *Jupiter's* orbit?
    → See on the **Solar System Data** chart on **RT 15** that Jupiter's eccentricity is **0.048**.
    → 0.345 is closer to 1 than 0.048 is. So your planet's orbit is **more eccentric** or **more elliptical** than Jupiter's orbit.

Earth Science Reference Tables Page 1

$$gradient = \frac{change\ in\ field\ value}{distance}$$

- **Gradient** = **slope** – a measurement of how much vertical change there is over an area
    - → The closer together the isolines, the steeper the gradient/slope.
    - → The further apart the isolines, the gentler the gradient/slope.
- **Change in field value** = value one minus value two (always a positive number)
- **Distance** = horizontal distance – measure using scale provided under the map
- Express your answer in **correct units**, such as *feet per mile*, or *meters per kilometer*.

Example: A 200-mile stream begins 800 feet above sea level and flows down to sea level (0 feet). What's the gradient of the stream?

→ $GRADIENT = \frac{800-0\ ft.}{200\ mi} = \mathbf{4\ ft/mi}$

$$rate\ of\ change = \frac{change\ in\ value}{time}$$

- **Rate of Change** – the speed at which a change takes place – how quickly a field value changes in a given amount of time
- **Change in value** – the field value – the area that's changing. **Ex** – temperature, pressure
    - → Change in field value must be positive (subtract smaller number from greater number)
- **Time** – change in time
- Express your answer in **correct units**, such as *degrees Celsius per minute*.

Example 1: The temperature of a rock changed from 40° C to 30° C between 2:10 and 2:13 p.m. What's the average rate of change expressed in degrees Celsius per minute?

→ $rate\ of\ change = \frac{change\ in\ value}{time} = \frac{40-30}{13-10} = \frac{10\ °C}{3\ min} = \mathbf{3.3°C/min.}$

Example 2: Between 5:00 AM to 8:00 AM, the temperature rose from 28 °C to 34 °C. What was the average rate of change expressed in degrees Celsius per hour?

→ $\frac{6°\ C}{3\ hours} = \mathbf{2°\ C/hr}$

Earth Science Reference Tables Page 1

$$\text{density} = \frac{mass}{volume}$$

- **Density** – how squashed together the particles of a substance are
- **Mass** – the amount of **matter** in a substance
- **Volume** – the amount of **space** a substance takes up
- Express your answer in **correct units**, such as *grams per cubic centimeter*
- Besides solving for density, also be able to **solve for mass or volume** given the density and one other part of equation.

<u>Example 1</u>: What is the *density* of a brick with a volume of 6 cm³ and a mass of 12 grams?
→ Formula: density = mass/volume → 12 g/6 cm³
→ Density = 2g/cm³

<u>Example 2</u>: What is the *volume* of a brick with density of 8g/cm³ and mass of 24 grams?
→ D = mass/volume → 8 = 24/V.
→ Volume = 3 cm³

1. The average monthly temperatures for Eureka, California, and Omaha, Nebraska, are plotted on the graph.

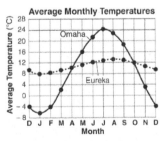

Calculate the rate of change in the average monthly temperature for Omaha during the two-month period between October and December, as shown on the graph.
___ °C/month

2. Point A represents a location on Earth's surface. Lines BC and XY are reference lines on the map. Points D, E, F, and G represent locations along Coe Creek. Elevations are shown in feet. Calculate the gradient along line XY. Label your answer with the correct units. _____

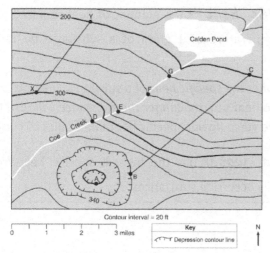

3. Cowlesville, New York, received a total of 88 inches of snow in 85 hours. Calculate the average rate of snowfall in inches per hour (in/h) for Cowlesville. _____ **in/h**

4. Base your answer to this question on the table below and on your knowledge of Earth science. The table shows the mineral characteristics of gold.

Mineral Characteristics of Gold

Luster	Hardness	Dominant Form of Breakage	Color	Streak	Density g/cm³	Chemical Symbol
metallic	2.5 to 3	fracture	golden yellow	golden yellow	19.3	Au

A gold nugget with a volume of 0.8 cubic centimeter (cm³) was found. Calculate the mass of this gold nugget. _____ **g**

Earth Science Reference Tables Page 1

Specific Heats of Common Minerals

Specific Heats of Common Materials	
MATERIAL	SPECIFIC HEAT (Joules/gram • °C)
Liquid water	4.18
Solid water (ice)	2.11
Water vapor	2.00
Dry air	1.01
Basalt	0.84
Granite	0.79
Iron	0.45
Copper	0.38
Lead	0.13

Specific heat → the amount of heat required to raise the temperature of 1 gram of a substance by 1 degree Celsius

The *higher the specific heat* of a substance (i.e. the longer it takes to heat up), the *longer the substance will take to cool* and vice versa (lower specific heat takes quick to cool).

Liquid water has the highest specific heat from all naturally occurring substances.

Reading the Table:

- The *left* column gives you the name of a **MATERIAL** commonly found on Earth.
    → Note that **water** is listed in all three phases – liquid, solid and vapor.
- The *right* column gives you the material's **SPECIFIC HEAT** in **JOULES/GRAM/°C**. *(Joule – a measure of heat energy.)*

Example 1: To raise one gram of copper from 9° C to 10° C, you need **0.38** joules of heat.

Example 2: To raise one gram of copper from 9° C to 13° C, you need 0.38 X 4 joules = **1.52** joules of heat. (0.38 joules per every degree Celsius).

5. Equal masses of basalt, granite, iron, and copper received the same amount of solar energy during the day. At night, which of these materials cooled down at the fastest rate?
   (1) basalt   (3) iron
   (2) granite   (4) copper

6. Which group of substances is arranged in order of decreasing specific heat values?
   (1) iron, granite, basalt   (3) dry air, water vapor, ice
   (2) copper, lead, iron   (4) liquid water, ice, water vapor

Earth Science Reference Tables Page 1

Properties of Water

This table gives you the amount of **heat energy** gained or *released* as water undergoes specific phase changes. It also gives you **water's maximum density.**

Properties of Water	
Heat energy gained during melting	334 J/g
Heat energy released during freezing	334 J/g
Heat energy gained during vaporization	2260 J/g
Heat energy released during condensation	2260 J/g
Density at 3.98°C	1.0 g/mL

Reading the Table:

- Water **gains** (absorbs) heat during **melting** and **vaporization** (evaporation). This means that these processes *require* heat.

- Water **releases** (loses) heat during **freezing** and **condensation**.

- Water's density at **3.98° C**, the temperature when it reaches its maximum possible density, is **1g/mL** (cm³).

FYI
*Most substances are densest as a solid. Water is the only naturally occurring substance that is densest as a liquid. As water vapor cools, it gets denser and turns into liquid water. As liquid water cools, it also gets denser – but only until it reaches 3.98° C. Then it reaches its **maximum density**. If you cool water at 3.98° C further, it will get colder, but less dense! That is why ice floats on liquid water – because it is less dense than water.*

7. Which change in the heat energy content of water occurs when water changes phase from a liquid to a solid?
   (1) gain of 334 Joules of heat energy per gram
   (2) release of 334 Joules of heat energy per gram
   (3) gain of 2260 Joules of heat energy per gram
   (4) release of 2260 Joules of heat energy per gram

8. During which phase change does water release the most heat energy?
   (1) freezing   (3) condensation
   (2) melting   (4) vaporization

9. The model below shows the movement of water in the water cycle.

How many joules of heat energy are required to evaporate 2 grams of water from this lake surface? _____ **joules**

Copyright 2019 © Y. Finkel | ALL RIGHTS RESERVED

Earth Science Reference Tables Page 1

Average Chemical Composition of Earth's Crust, Hydrosphere, and Troposphere

> This table shows which **elements** make up Earth's **crust** (outer layer of Earth), **hydrosphere** (ocean) and **troposphere** (layer of atmosphere closest to Earth) and their **percents** by **volume** or **mass**.

Average Chemical Composition of Earth's Crust, Hydrosphere, and Troposphere

ELEMENT (symbol)	CRUST		HYDROSPHERE	TROPOSPHERE
	Percent by mass	Percent by volume	Percent by volume	Percent by volume
Oxygen (O)	46.10	94.04	33.0	21.0
Silicon (Si)	28.20	0.88		
Aluminum (Al)	8.23	0.48		
Iron (Fe)	5.63	0.49		
Calcium (Ca)	4.15	1.18		
Sodium (Na)	2.36	1.11		
Magnesium (Mg)	2.33	0.33		
Potassium (K)	2.09	1.42		
Nitrogen (N)				78.0
Hydrogen (H)			66.0	
Other	0.91	0.07	1.0	1.0

Reading the Table:

♦ The column furthest to the *left* gives you the **names** and **chemical symbols** of different **ELEMENTS** that make up Earth.

♦ The next column, labeled **CRUST**, is divided into two. It tells you what percentage (by mass or volume) of each element on the left is found in Earth's crust. Percentages by mass are arranged from *greatest* to *least*.

Ex: In what percentage *by mass* is aluminum found in Earth's crust?
→ 8.23%

Ex: Which element is second most abundant by volume in Earth's crust?
→ Potassium

THE REFERENCE TABLES UNEARTHED | A Clear & Simple Reference Tables Guide

- The next two columns, **HYDROSPHERE & TROPOSPHERE**, tell you the **percentage by volume** of the elements that compose it.

Ex: Which two elements make up most of Earth's troposphere?

→ Oxygen & nitrogen

10. The pie graph below represents the composition, in percent by mass, of the chemical elements found in an Earth layer.

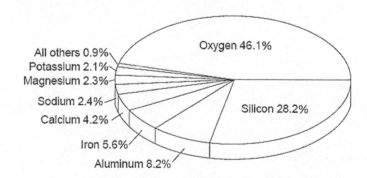

The composition of which Earth layer is represented by the pie graph?
(1) crust
(2) outer core
(3) troposphere
(4) hydrosphere

11. Which two elements make up the greatest percentages by mass in Earth's crust?
(1) oxygen and potassium
(2) oxygen and silicon
(3) aluminum and potassium
(4) aluminum and silicon

12. What is the approximate percent of oxygen by volume present in Earth's lower atmosphere?
(1) 21%
(2) 33%
(3) 46%
(4) 94%

13. In which two Earth regions is oxygen the second most abundant element by volume?
(1) crust and hydrosphere
(2) hydrosphere and troposphere
(3) troposphere and core
(4) core and crust

NEW YORK STATE
EARTH SCIENCE REFERENCE TABLES

PAGES 2-3

- *Generalized Landscape Regions of New York State*
- *Generalized Bedrock Geology of New York State*

Earth Science Reference Tables Pages 2-3

Generalized Landscape Regions of New York State

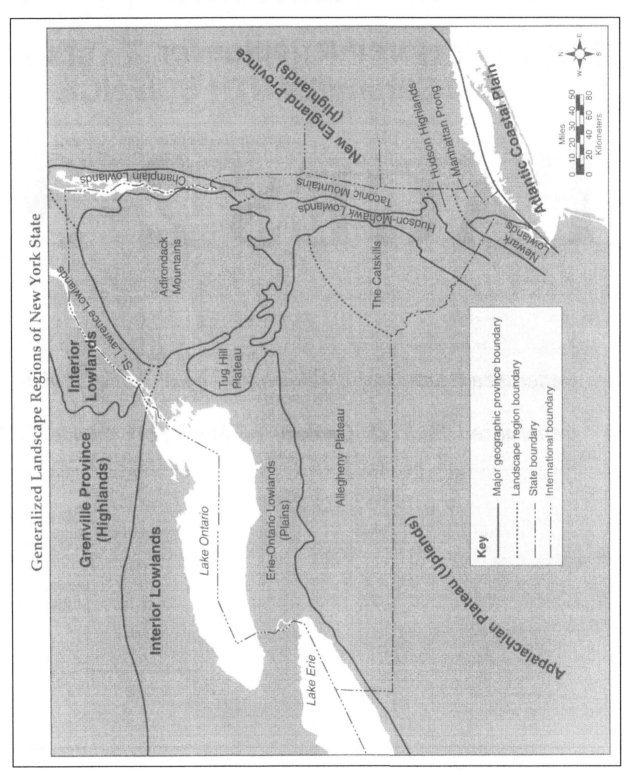

Generalized Bedrock Geology of New York State

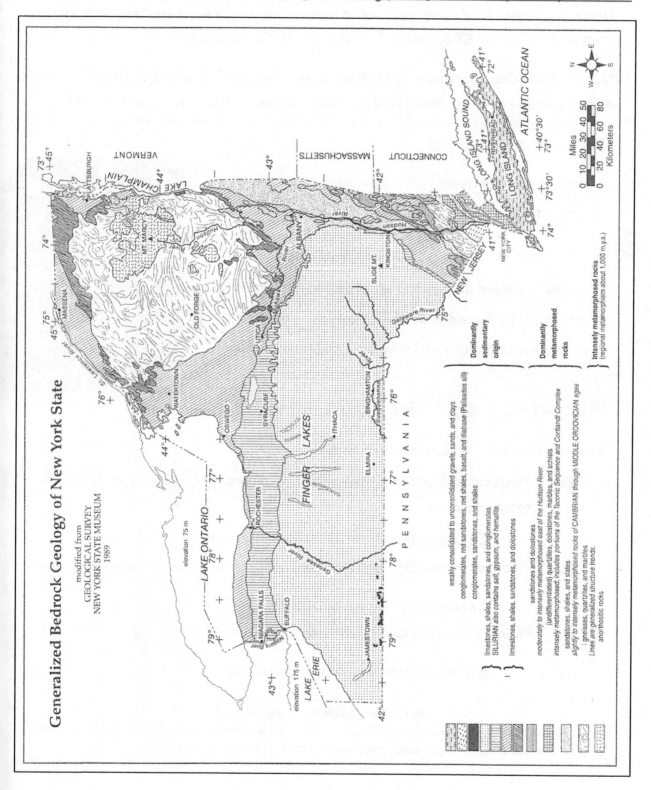

Earth Science Reference Tables Pages 2-3

> RT 2, **Generalized Landscape Regions of New York State**, is a *map of New York State*, divided into *landscape regions*. It corresponds to the *Generalized Bedrock Geology of New York State* map on **RT 3**, and the two maps are often used together.
>
> RT 3 – Bedrock Geology of New York State – is also a *map of New York State*. As stated above, it corresponds to the *Generalized Landscape Regions of New York State* map on **RT 2**.

RT 3 Includes:

- **Mountains** and **rivers** found in the state:
    → Note the locations of…
    - **Mt. Marcy** in the *Adirondack Mountains*
    - **Finger Lakes** in the *Allegheny Plateau*
- A **key** identifying the type and age of *bedrock* of each region
- The **latitude & longitude** of the locations on the map
    → see below for a discussion on latitude and longitude

Note: The *New England Province, Appalachian Plateau, Grenville Province*, and the *Interior Lowlands* north of Lake Ontario are *not part of New York State*. Page 3 shows you that these areas are part of Connecticut, Massachusetts, Vermont, Pennsylvania and Canada, respectively. Thus, any question about this map will not address these areas.

Reading the Tables:

- There are 3 major landscape regions on *RT 2*:

    1. *Mountains* (Highlands) – **high** elevation, composed of *distorted* (non-horizontal) *metamorphic* rock

    Ex: Adirondack Mountains, Hudson Highlands

    2. *Plateaus* (Highlands) – **high** elevation, composed of *horizontal sedimentary* layers

    Ex: Allegheny Plateau, The Catskills

    3. *Plains* (Lowlands) – **low** elevation, *sedimentary* rock

    Ex: Interior Lowlands, Atlantic Coastal Plain

The Catskill "Mountains" – a Misnomer:
Although the Catskill region is commonly known as "The (Catskill) Mountains," this is a misnomer. The Catskills region is a **plateau**. Proof – it is composed of sedimentary (and not metamorphic) rock.

THE REFERENCE TABLES UNEARTHED | A Clear & Simple Reference Tables Guide

- The KEY on the bottom left of **RT 3** tells you:
  - → the names of the **specific rocks** that compose the bedrock
  - → the **general makeup** of the bedrock.

> **Mountains vs. Plateaus**
> To differentiate between a mountain & a plateau when the reference tables just calls the region a "Highland," **check its bedrock composition.** If *metamorphic*, it's a mountain (Hudson Highlands) and if *sedimentary*, it's a plateau (The Catskills).

Example 1: **For properties of the Adirondack Mountains:**
- → Find the Adirondack Mountains on **RT 2**
- → Match the area to the map on **RT 3** (Notice cities *Old Forge* and *Mt. Marcy* included in this region)
- → Notice the pattern of this area, and find this pattern on the bottom left of **RT 3**
- → See that the bedrock of the Adirondack Mountains is composed of **gneisses, quartzites** and **marbles**, and is made up of **intensely metamorphosed rock**.

Example 2: **For properties of the Tug Hill Plateau:**
- → Find the *Tug Hill Plateau* on **RT 2**
- → Match the area to the map on **RT 3** (Notice the cities *Watertown* and *Oswego* – they are **not** included in this region – they are part of *Erie-Ontario Lowlands*!)
- → Notice the pattern of this area, and find this pattern on the bottom left of **RT 3**
- → See that the bedrock of the Tug Hill Plateau is composed of **limestones, shales, sandstones** and **conglomerates**, and is made up of **sedimentary rock**.

Earth Science Reference Tables Pages 2-3

Latitude & Longitude

- When giving the coordinates of a location, **latitude** goes first, then **longitude**

- The little numbers on the **right** and **left** edges of the map on **RT 3** show the *latitude* (TIP: Use a **straightedge** from one edge of the map to the other to calculate *exact* latitude)
  → Each degree is *subdivided* into **60 minutes (')**
  Ex: Latitude of *Ithaca* is approximately **42°30' N**
  Ex: Latitude of *Mt. Marcy* is approximately **44°05' N**
  → Latitude in New York State ranges from **40°30' N** to **45° N**

- The little numbers on the **top** and **bottom** edges of the map on **RT 3** show the *longitude*
  → Longitude in New York State ranges from **72°** to **79°45'**
  Ex: Longitude of *Ithaca* is **76°30' W**
  Ex: Longitude of *Mt. Marcy* is **73°55' W**

> **HELPFUL TIP:**
> Since New York State is in the northwestern hemisphere, any latitude in New York State is NORTH & any longitude in New York state is WEST.

Polaris & Latitude

- **Polaris** = the North Star – a star located *directly above the North Pole*

- **Altitude to Polaris** = the angle formed between the (imaginary) line to horizon and line to Polaris.

- **Altitude to Polaris = degree of latitude!** (and vice versa)

  Ex: Altitude to Polaris in *Watertown* is **44°** because its *latitude* is 44° N

  Ex: Altitude to Polaris in *Elmira* is **42°10'** because its *latitude* is 42°10' N

> **More Practice:**
> What's the latitude and longitude of...
> 1. Watertown?
>   44° N, 75°55' W
> 2. Elmira?
>   42°10' N, 76°50' W
> 3. Albany?
>   42°40' N, 73°45' W
> 4. Buffalo?
>   42°50' N, 78°50' W

14. In which landscape region are New York State's Finger Lakes primarily located?

    (1) Adirondack Mountains
    (2) Allegheny Plateau
    (3) Atlantic Coastal Plain
    (4) Erie-Ontario Lowlands

15. The Adirondacks are classified as mountains because of the high elevation and bedrock that consists mainly of

    (1) deformed and intensely metamorphosed rocks
    (2) glacial deposits of unconsolidated gravels, sands, and clays
    (3) quartzites and marbles
    (4) horizontal sedimentary rocks of marine origin

16. The surface bedrock of Mt. Marcy, New York, is composed primarily of which rock?

    (1) anorthosite
    (2) marble
    (3) quartzite
    (4) hornfels

17. Which New York State landscape region is composed of mostly horizontal sedimentary bedrock and has a high elevation?

    (1) Hudson Highlands
    (2) Manhattan Prong
    (3) the Catskills
    (4) Taconic Mountains

18. Which chart best describes the landscape category and the general bedrock structure, type, and composition of New York State's Catskills?

    (1)
Landscape Category	plateau
Bedrock Structure	horizontal
Bedrock Type	sedimentary
Bedrock Composition	limestone, shale, sandstone

    (2)
Landscape Category	mountain
Bedrock Structure	folded
Bedrock Type	sedimentary
Bedrock Composition	sandstone, dolostone, schist

    (3)
Landscape Category	mountain
Bedrock Structure	horizontal
Bedrock Type	metamorphic
Bedrock Composition	gneiss, quartzite, marble

    (4)
Landscape Category	plateau
Bedrock Structure	folded
Bedrock Type	metamorphic
Bedrock Composition	shale, slate, dunite

Earth Science Reference Tables Pages 2-3

19. Base your answer to this question on the map of Mendon Ponds Park below and on your knowledge of Earth science.

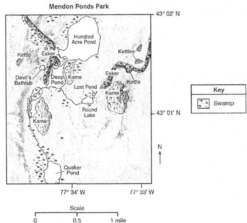

Based on the map, in which New York State landscape region is Mendon Ponds Park located?

(1) Allegheny Plateau
(2) Tug Hill Plateau
(3) Erie-Ontario Lowlands
(4) Hudson-Mohawk Lowlands

Base your answers to questions 21 and 22 on the map below and on your knowledge of Earth science. The map shows the locations of major watersheds in New York State. Letters A through K represent individual watersheds.

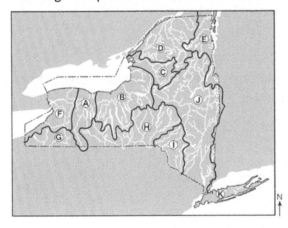

20. In which major watershed is the Susquehanna River located?

(1) F
(2) H
(3) I
(4) J

21. Over which two landscape regions do the streams in watershed D flow?

(1) Tug Hill Plateau and the Catskills
(2) Tug Hill Plateau and Erie-Ontario Lowlands
(3) Adirondack Mountains and Champlain Lowlands
(4) Adirondack Mountains and St. Lawrence Lowland

NEW YORK STATE
EARTH SCIENCE REFERENCE TABLES

PAGE 4

- *Surface Ocean Currents*

Earth Science Reference Tables Page 4

Surface Ocean Currents

This table shows you the paths of the major **surface ocean currents** on Earth.
Surface ocean currents – the continuous flow of water along the ocean's surface.

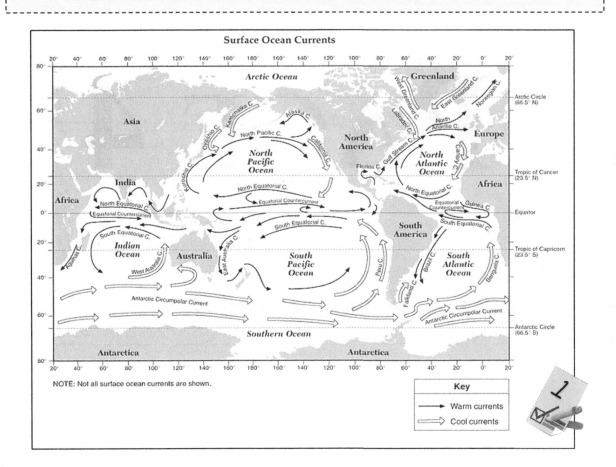

Reading the Table:

- The **KEY** on the bottom right gives you the **temperature** of the currents:

  → The **black arrows** represent warm currents, which cause **warm** and **moist** climates

  → The **outlined white arrows** represent cool currents, which cause **cool** and **dry** climates

- Most questions ask about a **current** and a **location**.

Example: Which cool current influences the west coast of North America?

- → Find the west coast of North America
- → Find a cool current – a current represented by an outlined arrow
- → **California** Current

♦ Notice the **latitude & longitude** markings on the sides of the map. The *left* gives you general latitude and the *right* gives locations of special latitudes, such the *Tropic of Cancer – 23.5° N*. The *top* and *bottom* give you the longitude.

- → Note: **0° longitude is not in the middle!** Any longitude to the *left* of 180° is **east** – and any longitude to the *right* of 180° is **west**.
  - ✓ HINT: Look at the continents – you [should!] know that **Asia & Australia** are in the *eastern hemisphere*.
  - ✓ **Ex:** The "**A**" from the word "Australia" is approximately 25° S 120° **E**, while the "**N**" from "North America" is approximately 45° N 100° **W**.

Direction of Surface Ocean Current Flow:

A combination of the following factors influences the direction of surface ocean current flow:

1. *Prevailing winds* – before other factors are involved, the currents will flow in roughly the same direction as prevailing winds
   - → See **RT 14 – Planetary Wind & Moisture Belts in the Troposphere** for a more in-depth discussion on prevailing winds
     - ✓ **Ex:** The **North Pacific Ocean Current**, located in the 30-60° wind belt, is moving [somewhat] towards the *northeast* – the same direction as the prevailing winds.

2. *Large landmasses* – currents will change direction when they hit **large landmasses**
   - ✓ **Ex:** The **South Pacific Ocean Current** changes direction when it hits *the west coast of South America*

3. *Coriolis effect* – currents will generally move **clockwise** in the Northern hemisphere and **counterclockwise** in the Southern hemisphere.
   - ✓ **Ex:** The **South Equatorial Current** is moving *counterclockwise*, while the **North Equatorial Current** is moving *clockwise*.

Earth Science Reference Tables Page 4

22. Which ocean current brings warm water to the southeastern coast of Africa?
    (1) Agulhas Current
    (2) West Australian Current
    (3) Benguela Current
    (4) Equatorial Countercurrent

23. Which current has a cooling effect on the climate of the west coast of South America?
    (1) Falkland Current
    (2) Peru Current
    (3) Benguela Current
    (4) Brazil Current

24. Which ocean current directly warms Western Europe?
    (1) North Atlantic Current
    (2) South Equatorial Current
    (3) Canary Current
    (4) Labrador Current

25. Base your answer to this question on the map below and on your knowledge of Earth science. The map indicates the locations of Eureka, California and Omaha, Nebraska. Identify the surface ocean current that affects the climate of Eureka.

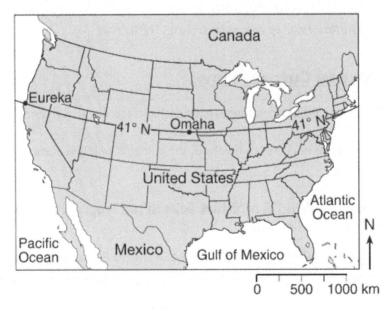

_____ current

FYI: The **Antarctic Circumpolar Current** is the only current that travels around the entire globe! (Landmasses block the other currents from doing the same).

NEW YORK STATE
EARTH SCIENCE REFERENCE TABLES

PAGE 5

- Tectonic Plates

Earth Science Reference Tables Page 5

Tectonic Plates

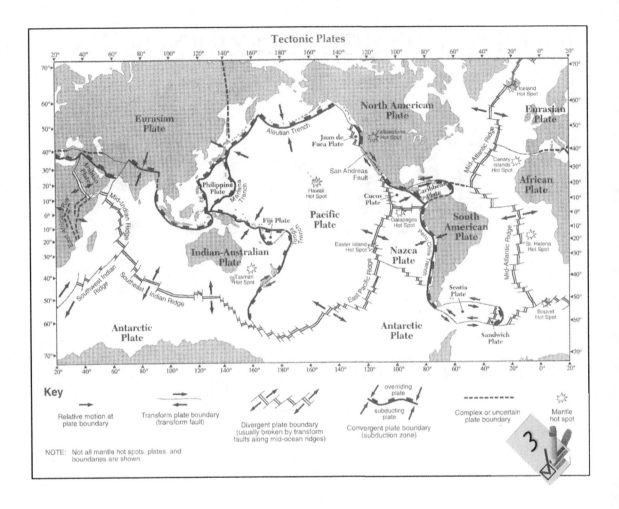

This table shows major **TECTONIC PLATES** that make up Earth's crust.

Tectonic plate – a large piece of Earth's lithosphere (outer layer of Earth)

It also shows:

→ the **types of boundaries** that exist along each plate

→ **tectonic features** such as **hot spots**, **mid-ocean ridges** and **trenches**

Reading the Table:

- Note the **KEY** on the bottom, indicating the type of boundary along the border of each plate

  → A **dark arrow** shows you the direction of plate movement.

  → *Transform plate boundary* – the plates *slide past each other*. This usually causes earthquakes.

  → *Divergent plate boundary* – the plates *move away from each other*. Notice **mid-ocean ridges** in those areas – formed when the plates shift and the empty space fills with magma

  → *Convergent plate boundary* – the plates *move towards each other*. Notice **trenches** in those areas– formed when the **subducting** plate (denser plate) sinks below the **overriding** plate (less dense plate).

  → *Complex or uncertain plate boundary* – it is unclear to scientists what kind of boundary exists in those areas.

  → *Mantle hot spot* – an area where there is a **hot spot** underneath in Earth's mantle – *hot rising magma* that forms volcanoes on Earth's crust directly above the area.

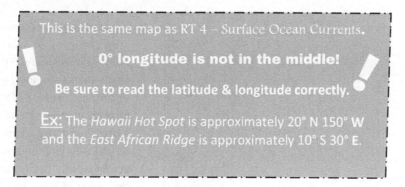

This is the same map as RT 4 – Surface Ocean Currents.

0° longitude is not in the middle!

Be sure to read the latitude & longitude correctly.

<u>Ex:</u> The *Hawaii Hot Spot* is approximately 20° N 150° W and the *East African Ridge* is approximately 10° S 30° E.

<u>Example 1:</u> What kind of tectonic plate movement exists between the Fiji Plate and the Indian-Australian Plate?

  → Converging plates/subduction zone

<u>Example 2:</u> Which tectonic feature formed the volcanic islands southeast of Australia?

  → Tasman Hot Spot

Earth Science Reference Tables Page 5

26. Base your answer to this question on the passage and the map of South America below and on your knowledge of Earth science.

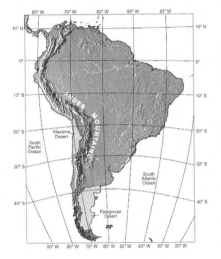

Two South American Deserts
South America is an excellent example of the influence that plate tectonic features have on climates. The Andes mountain range, formed by plate tectonics, is on the western edge of South America.

Name one tectonic plate that is interacting with the South American Plate to uplift the Andes Mountains. _____ **plate**

27. Base your answer to this question on the passage and map below and on your knowledge of Earth science. The passage describes the March 11, 2011, earthquake that occurred off the coast of Japan and the tsunami that it generated. The map shows the location of the earthquake epicenter and the tsunami travel times across the Pacific Ocean.

Earthquake & Tsunami Rattle the Pacific
At 2:46 p.m., in Japan, on Friday, March 11, 2011, a magnitude 9.0 earthquake occurred below the ocean floor, at a depth of 18.6 miles under the ocean surface. The epicenter was located approximately 80 miles off Japan's eastern coast at the approximate coordinates of 38° N 142° E. The earthquake shook buildings across Japan. The tsunami also occurred along the coasts of other countries and islands in the Pacific Ocean.

Identify the type of plate boundary where the earthquake occurred. _____ **plate**

28. Which surface feature was produced by crustal movements at a transform plate boundary?
   (1) African Rift
   (2) Aleutian Trench
   (3) Tasman Hot Spot
   (4) San Andreas Fault

29. Which type of tectonic plate boundary is found between the South American Plate and the Scotia Plate?
    (1) transform
    (2) convergent
    (3) divergent
    (4) complex/uncertain

30. The photograph below shows the East African Rift Valley in Africa. Which tectonic movement of Earth's crust is most likely responsible for this feature?
    (1) convergence of continental crust
    (2) convergence of oceanic crust
    (3) divergence of continental crust
    (4) divergence of oceanic crust

31. Oceanic crust is sliding beneath the Aleutian Islands in the North Pacific Ocean, forming the Aleutian Trench at a
    (1) convergent plate boundary between the Pacific Plate and the North American Plate
    (2) convergent plate boundary between the Pacific Plate and the Juan de Fuca Plate
    (3) divergent plate boundary between the Pacific Plate and the North American Plate
    (4) divergent plate boundary between the Pacific Plate and the Juan de Fuca Plate

32. Which two features are commonly found at divergent plate boundaries?
    (1) mid ocean ridges and rift valleys
    (2) wide valleys and deltas
    (3) ocean trenches and subduction zones
    (4) hot spots and island arcs

33. Base your answer to this question on the cross sections below and on your knowledge of Earth science. The cross sections represent three different stages in the development of Denali (Mt. McKinley) and the growth of the North American Plate in Alaska near the boundary with the Pacific Plate. Arrows represent the direction of plate movement.

    Identify the type of plate boundary represented in the cross sections.
    _____ plate

NEW YORK STATE
EARTH SCIENCE REFERENCE TABLES

PAGE 6

- Relationship of Transported Particles to Water Velocity
- Rock Cycle in Earth's Crust
- Scheme for Igneous Rocks Identification

Earth Science Reference Tables Page 6

Relationship of Transported Particle Size to Water Velocity

This table shows the **stream velocities** (speeds) necessary to transport different sized particles.

The *greater the stream velocity*, the **larger** and the **more** particles the stream could carry.

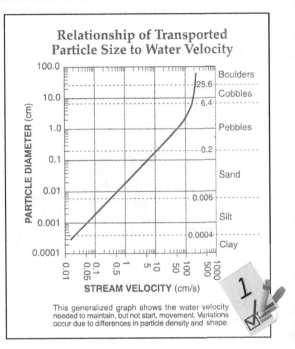

When using this table, use a straight edged item, such as a ruler or a piece of paper, since the lines are very small and easily confused.

Reading the Table:

- The left side gives **PARTICLE DIAMETER** in **centimeters**. The sizes correspond to the **sediment names** on the right side.
    - ➔ **Ex:** *Clay* ranges from **0.0001 cm** to **0.0004 cm** in diameter
    - ➔ **Ex:** *Silt* ranges from **0.0004 cm** to **0.006 cm** in diameter
- The bottom of the graph gives the **STREAM VELOCITY** in **centimeters per second**.
- At any specific velocity, the stream can carry only the sediments *below the* **black line** (because those are small enough for the stream moving at that velocity to carry them).
- If the stream velocity drops, the stream will **deposit** any particles that it can no longer carry.
    - ➔ **Ex:** A stream moving at a rate of **1 cm/s** can carry *clay, silt,* **and** *some sand*.
    - ➔ **Ex:** A stream moving at a rate of 100 cm/s can carry *clay, silt, sand* and *some pebbles*.
    - ➔ **Ex:** If a stream is moving at a rate of **10 cm/s** and then *its velocity drops to* **1 cm/s**, it will deposit *some sand particles*.

34. What is the approximate minimum stream velocity needed to keep a particle in motion that has a diameter of 10 centimeters?
    (1) 110 cm/s
    (2) 190 cm's
    (3) 325 cm/s
    (4) 425 cm/s

35. *A river's current carries sediments into the ocean. Which sediment size will most likely be deposited in deeper water farthest from the shore?
    (1) pebble
    (2) sand
    (3) silt
    (4) clay

Base your answers to questions 37 and 38 on the cross section and data table below and on your knowledge of Earth science.

The cross section shows the profile of a stream that is flowing down a valley from its source. Points A through E represent locations in the stream. The data table shows the average stream velocity at each location. The volume of water in the stream remains the same at all locations.

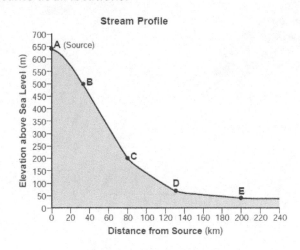

Location in Stream	Average Stream Velocity (cm/s)
A	10
B	110
C	130
D	20
E	15

Questions marked with an asterisk (*), such as question 37, are based on background information necessary for understanding the reference tables, but not based on information found in the tables themselves.

36. *The average stream velocity at each location is controlled primarily by the
    (1) elevation above sea level
    (2) slope of the land
    (3) sediment carried by the stream
    (4) distance from the stream's source

37. What is the largest type of sediment that could be transported at location B?
    (1) silt
    (2) pebbles
    (3) sand
    (4) cobbles

Earth Science Reference Tables Page 6

Base your answers to questions 39 and 40 on the passage and two diagrams below, and on your knowledge of Earth science.

The passage describes a method used to mine gold and the diagrams represent two different views of a sluice box, which is used to separate gold from other sediments.

Gold Mining

A sluice box is used to remove gold pieces from other sediments in a stream. The box is placed in the stream to channel some of the water flow. Gold-bearing sediment is placed at the upper end of the box. The riffles in the bottom of the box are designed and positioned to create disruptions in the water flow. These disruptions cause dead zones in the current that allow the more dense gold to drop out of suspension and be deposited behind the riffles. Lighter material flows out of the box as tailings. Typically, particles of the mineral pyrite, which shares characteristics with gold, are deposited with gold particles in the sluice box. Since miners were fooled into thinking the nuggets of pyrite were gold, the name "fool's gold" is often applied to pyrite.

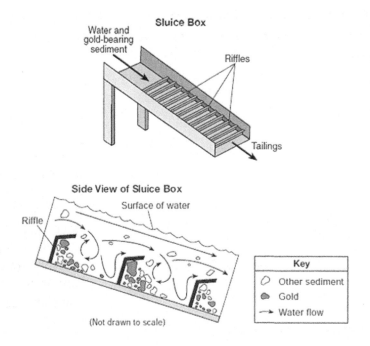

38. The velocity of the water leaving the sluice box was 90 centimeters per second (cm/s). State the diameter of the largest particle that could be found in the tailings. _____ cm

39. *The angle of the sluice box is changed so that the box has a steeper slope. Describe the most likely change in water velocity and the amount of sediment passing through the sluice box as tailings.
Water velocity: _____
Amount of sediment: _____

Rock Cycle in Earth's Crust

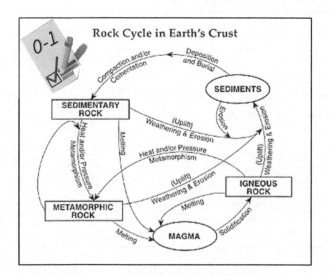

This table shows how the three rock types, *sedimentary, igneous* and *metamorphic*, are **interrelated**.

The table also shows the **processes** that produce each rock type.

<u>Note:</u> There is no preferred order for reading this table. Any kind of rock can become any other kind of rock if the right processes occur.

Reading the Table:

- Starting from the **SEDIMENTARY ROCK** box, you can see that if the sedimentary rock undergoes...

    → **heat/pressure** → a *metamorphic* rock

    → **melting** → magma → solidifies into an *igneous* rock

    → **weathering & erosion** → sediments → deposition/burial → compaction / cementation (gets pressed together) → new *sedimentary* rock.

- If the **METAMORPHIC ROCK** undergoes...

    → **Melting** → magma → solidifies into an *igneous* rock

    → **Weathering & erosion** → sediments → deposition & burial → compaction/cementation → *sedimentary* rock

    → **Heat &/or pressure** → new *metamorphic* rock

- If the **IGNEOUS ROCK** undergoes...

    → **Heat &/or pressure** → *metamorphic* rock

    → **Melting** → magma → solidifies into a new *igneous* rock

    → **Weathering & erosion** → sediments → deposition & burial → compaction/cementation → *sedimentary* rock

Earth Science Reference Tables Page 6

Scheme for Igneous Rock Identification

> This table helps you **identify** different **igneous rocks** based on specific properties you are given.
>
> **Igneous rock** – a rock formed by the solidification of molten magma

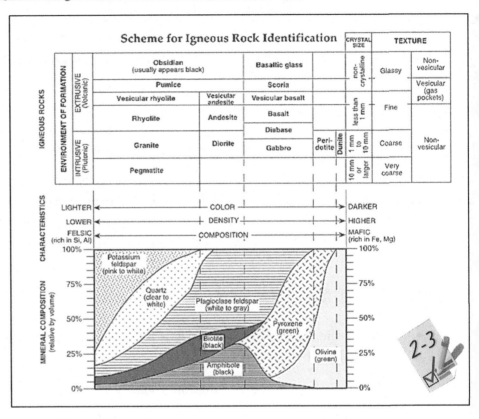

Reading the Table:

- The *top middle* of the chart gives you the names of various kinds of igneous rocks that you are identifying **in bold**. The rest of the chart refers to these rocks.

- The top left tells you the **ENVIRONMENT OF FORMATION** – whether the igneous rock is **EXTRUSIVE (volcanic)** – it formed *on or above* Earth's surface – or **INTRUSIVE (plutonic)** – it formed *below* Earth's surface.

    → All the rocks on the *top half* of the chart, such as ***Obsidian, Pumice, Basaltic glass & Andesite,*** are **extrusive**.

    → All the rocks on the *bottom half* of the chart, such as ***Granite, Gabbro & Dunite,*** are **intrusive**.

    → ***Diabase,*** which is on the *dividing line* between extrusive and intrusive, can be **either extrusive or intrusive**.

- **CRYSTAL SIZE** and **TEXTURE** on the *right* describes the igneous rocks' **crystalline structure** & the rocks' **texture** based on this crystalline structure:

- → When molten magma/lava solidifies to form igneous rocks, it usually solidifies in a process called **crystallization** – forms crystals (like rock candy).

- → The **size** of the crystals depends on the **speed** at which the rock cools.
    - ✓ **Some extrusive rocks** cool so *quickly* that there is no time for crystals to form. Then the rock will be **non-crystalline** and have a **glassy** texture.
    - ✓ **Other extrusive rocks** cool relatively *quickly*, but there is still time for tiny crystals, **less than 1 mm** in size, to form. These rocks have a **fine** texture.
    - ✓ **Intrusive rocks** cool more *slowly*. Depending on how slowly they cool, their crystals will be either from **1 mm to 10 mm** or **10 mm or larger** in size with **coarse** & **very coarse** textures respectively.

- → As the rocks cool, they sometimes trap bubbles of gas. This gives them a **vesicular** (gas pocket) texture.
    - ✓ Some *glassy* rocks, such as **Obsidian & Basaltic glass**, have a **non-vesicular** texture.
    - ✓ Other *glassy* rocks, such as **Scoria & Pumice**, and some fine rocks, such as **Vesicular Basalt**, have a **vesicular** texture.
    - ✓ The *rest of the rocks* on the chart – from *fine* to *very course,* have a **non-vesicular texture.**

♦ The **CHARACTERISTICS** row in middle of the chart describes the **color, density & composition** of the rocks named above.

- → Rocks towards the *left,* such as **pumice & granite**, have a **lighter** color, **lower** density and a **felsic** (rich in Si – silicon – and Al – aluminum) **composition**
- → Rocks towards the *right,* such as **gabbro & basaltic glass**, have a **darker** color, **higher** density and a **mafic** (rich in Fe – iron- and Mg – magnesium) **composition**.

♦ The **MINERAL COMPOSITION** chart on the bottom gives you the **percent by volume** of each mineral found in the above rocks. (Also gives **colors** of each mineral.)

- → **Ex:** The rocks **rhyolite & pegmatite** are composed of approximately *30% potassium feldspar, 20% quartz, 20% plagioclase feldspar, 15% biotite* and *15% amphibole.* [Note – these are just rough estimates – you do not need to calculate exact amounts.]
- → **Ex: Diorite** is composed of 2% quartz, 68% plagioclase feldspar, 10% biotite and 20% amphibole.

Earth Science Reference Tables Page 6

- **Putting It All Together:** Which igneous rock is composed of mostly pyroxene & plagioclase feldspar, has a fine, non-vesicular texture and is always extrusively formed?
    - → *"mostly pyroxene & plagioclase feldspar"* → basaltic glass, scoria, vesicular basalt, basalt, diabase & gabbro
    - → *"fine, non-vesicular texture"* → ~~basaltic glass~~, ~~scoria~~, ~~vesicular basalt~~, basalt, diabase, gabbro
    - → *"is always extrusively formed"* → **basalt**, ~~diabase~~, ~~gabbro~~

40. An igneous rock has mineral crystals ranging in size from 2 to 6 millimeters. The rock is composed of 58% plagioclase feldspar, 26% amphibole, and 16% biotite. What is the name of this rock?
    (1) diorite
    (2) gabbro
    (3) andesite
    (4) pumice

41. The igneous rock gabbro most likely formed from molten material that cooled
    (1) rapidly at Earth's surface
    (2) slowly at Earth's surface
    (3) rapidly, deep underground
    (4) slowly, deep underground

42. Which graph best shows the relationship between the compositions of different igneous rocks and their densities?

(1)　　　　　(2)　　　　　(3)　　　　　(4)

43. Basaltic rock that forms volcanic mountains where mantle plumes reach Earth's surface is usually composed of
    (1) fine-grained, dark-colored felsic minerals
    (2) fine-grained, dark-colored mafic minerals
    (3) coarse-grained, light-colored felsic minerals
    (4) coarse-grained, light-colored mafic minerals

44. Base your answer to this question on the passage below and on your knowledge of Earth science.

Island Arcs

Island arcs are long, curved chains of oceanic islands associated with seismic activity and mountain-building processes at certain plate boundaries. They occur where oceanic tectonic plates collide. Along one side of these island arcs, there is usually a long, narrow deep-sea trench.

At island arcs, the denser plate is subducted and is forced into the partially molten mantle under the less dense plate. The islands are composed of the extrusive igneous rocks basalt and andesite. The basalt originates most likely from the plastic mantle. The andesite originates most likely from the melting of parts of the descending plate and sediments that had accumulated on its surface.

Which list identifies minerals present in andesite from the greatest percentage by volume to the least percentage by volume?
(1) biotite, plagioclase feldspar, amphibole
(2) biotite, amphibole, plagioclase feldspar
(3) plagioclase feldspar, biotite, amphibole
(4) plagioclase feldspar, amphibole, biotite

45. Base your answer to this question on the geologic cross section and graph below, and on your knowledge of Earth science. The cross section represents the intrusive igneous rock of the Palisades sill and surrounding bedrock located on the west side of the Hudson River across from New York City. The graph indicates changes in the percentages of the major minerals found in the sill.

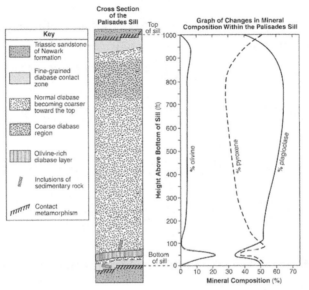

Which two minerals, not shown on the Graph of Changes in Mineral Composition Within the Palisades Sill, are also likely to be found in some other samples of diabase?

(1) quartz and biotite
(2) biotite and amphibole
(3) amphibole and potassium feldspar
(4) potassium feldspar and quartz

46. Base your answer to this question on the photographs below and on your knowledge of Earth science. The photographs show eight common rock-forming minerals. In the table below, place an X in the appropriate box to indicate whether each mineral is found mainly in felsic or mafic igneous rock.

Mineral Name	Felsic	Mafic
Potassium feldspar		
Olivine		
Quartz		
Pyroxene		

47. Base your answer to this question on the cross sections below and on your knowledge of Earth science. The cross sections represent three different stages in the development of Denali (Mt. McKinley) and the growth of the North American Plate in Alaska near the boundary with the Pacific Plate. Arrows represent the direction of plate movement.

Formation of Denali (Mt. McKinley)

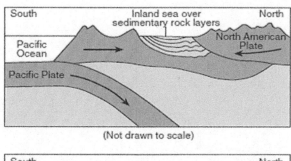

Stage 1
Years Ago:
Sedimentary rocks that would later form Denali's (Mt. McKinley's) north peak began as sediments deposited under an inland sea.

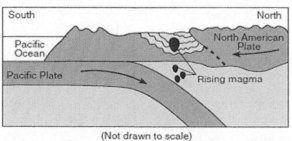

Stage 2
Years Ago:
Magma rose into the sedimentary rocks. This would later form the granite rock making up Denali's (Mt. McKinley's) south peak. Tectonic forces continued to push up the land surface.

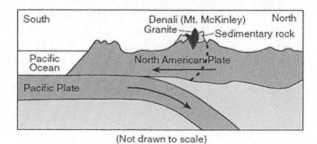

Stage 3
Today:
Tectonic forces continue to cause uplift in the region.

Circle either "volcanic" or "plutonic" below to identify the environment of formation of the granite found on Denali (Mt. McKinley). Describe the cooling rate of the magma that produced this granite.

Circle one:
volcanic   plutonic
Cooling rate: _____

48. Andesite makes up much of the volcanic rock of the Andes Mountains. Name three minerals that are commonly found in a single andesite rock.

_____   _____   _____

NEW YORK STATE
EARTH SCIENCE REFERENCE TABLES

PAGE 7

- *Scheme for Sedimentary Rock Identification*
- *Scheme for Metamorphic Rock Identification*

Scheme for Sedimentary Rock Identification

Earth Science Reference Tables Page 7

3 ✓ On the following two tables

Scheme for Sedimentary Rock Identification

INORGANIC LAND-DERIVED SEDIMENTARY ROCKS

TEXTURE	GRAIN SIZE	COMPOSITION	COMMENTS	ROCK NAME	MAP SYMBOL
Clastic (fragmental)	Pebbles, cobbles, and/or boulders embedded in sand, silt, and/or clay	Mostly quartz, feldspar, and clay minerals; may contain fragments of other rocks and minerals	Rounded fragments	Conglomerate	
			Angular fragments	Breccia	
	Sand (0.006 to 0.2 cm)		Fine to coarse	Sandstone	
	Silt (0.0004 to 0.006 cm)		Very fine grain	Siltstone	
	Clay (less than 0.0004 cm)		Compact; may split easily	Shale	

CHEMICALLY AND/OR ORGANICALLY FORMED SEDIMENTARY ROCKS

TEXTURE	GRAIN SIZE	COMPOSITION	COMMENTS	ROCK NAME	MAP SYMBOL
Crystalline	Fine to coarse crystals	Halite	Crystals from chemical precipitates and evaporites	Rock salt	
		Gypsum		Rock gypsum	
		Dolomite		Dolostone	
Crystalline or bioclastic	Microscopic to very coarse	Calcite	Precipitates of biologic origin or cemented shell fragments	Limestone	
Bioclastic		Carbon	Compacted plant remains	Bituminous coal	

This table helps you **identify** different **sedimentary rocks** based on specific properties.

Sedimentary rock – a rock formed by the accumulation of sediments (pieces of pre-existing rock) and/or organic material (remains of living things)

Reading the Table:

- The chart is split into 2 parts:

    → **INORGANIC LAND-DERIVED SEDIMENTARY ROCKS** – rocks formed on *land* just from *sediments* of other rocks *without any organic material*

    → **CHEMICALLY AND/OR ORGANICALLY FORMED SEDIMENTARY ROCKS** – rocks formed by *chemical* means and/or *with organic materials*.

- The two columns to the far right, **ROCK NAME** and **MAP SYMBOL**, give you the names and symbols of the sedimentary rocks that this chart is discussing.

    → **The symbols are used very often on diagrams on regents questions on other topics – you must know to look at this chart to identify which rock is which.**

- The **TEXTURE** column on the *far left* describes the texture of the sedimentary rocks:

  → **Clastic (fragmental)** – composed of pieces called *sediments/clasts* (usually rounded because moved by water or wind)

  → **Crystalline** – rocks formed with a *crystal structure* when dissolved minerals precipitate (fall out) from solutions (because of evaporation, saturation with dissolved minerals or changes in temperature)

  → **Crystalline or bioclastic** – see above & below

  → **Bioclastic** – rocks made by *living organisms* or composed of *organic materials* (**ex:** clam's shell)

- The **GRAIN SIZE** column describes the grain size of the sedimentary rocks:

  → In **inorganic rocks**, grain size ranges from *pebbles/cobbles/boulders* to *clay*. (See **RT 6 – Relationship of Transported Particles to Stream Velocity** for all sediments and their sizes.)

  → In **chemical/organic rocks**, grain size ranges from *fine to coarse crystals* in **crystalline rocks** and from *microscopic to very coarse* in **bioclastic and some crystalline rocks**.

- The **COMPOSITION** column describes the mineral composition of the sedimentary rocks.

  → The **clastic rocks** are all composed mostly of *quartz, feldspar & clay*.

  → The **crystalline/bioclastic rocks** each have a *specific mineral composition*.

- The **COMMENTS** column describes different properties specific to each kind of sedimentary rock not yet discussed on the table.

  → **Ex:** Siltstone is composed of *very fine grain*

  → **Ex:** Bituminous coal is composed of *compacted* (pressed together) *plant remains*

- <u>**Putting It All Together:**</u> Which sedimentary rock has a *crystalline texture* and is *composed of gypsum*?

  → "crystalline texture" → rock salt, rock gypsum, dolostone

  → "composed of gypsum" → ~~rock salt,~~ **rock gypsum,** ~~dolostone~~

Earth Science Reference Tables Page 7

Scheme for Metamorphic Rock Identification

Scheme for Metamorphic Rock Identification

TEXTURE		GRAIN SIZE	COMPOSITION	TYPE OF METAMORPHISM	COMMENTS	ROCK NAME	MAP SYMBOL
FOLIATED	MINERAL ALIGNMENT	Fine	MICA QUARTZ FELDSPAR AMPHIBOLE GARNET PYROXENE	Regional (Heat and pressure increases)	Low-grade metamorphism of shale	Slate	
		Fine to medium			Foliation surfaces shiny from microscopic mica crystals	Phyllite	
					Platy mica crystals visible from metamorphism of clay or feldspars	Schist	
	BANDING	Medium to coarse			High-grade metamorphism; mineral types segregated into bands	Gneiss	
NONFOLIATED		Fine	Carbon	Regional	Metamorphism of bituminous coal	Anthracite coal	
		Fine	Various minerals	Contact (heat)	Various rocks changed by heat from nearby magma/lava	Hornfels	
		Fine to coarse	Quartz	Regional or contact	Metamorphism of quartz sandstone	Quartzite	
			Calcite and/or dolomite		Metamorphism of limestone or dolostone	Marble	
		Coarse	Various minerals		Pebbles may be distorted or stretched	Metaconglomerate	

This table helps you identify different metamorphic rocks based on specific properties.

Metamorphic rock – a rock formed from **changes in previously existing rocks** due to *heat, pressure* and/or *mineral fluids* without weathering or melting

Metamorphism – the process of forming metamorphic rocks

Reading the Table:

- The two columns to the *far right*, **ROCK NAME** and **MAP SYMBOL**, give you the names and symbols of the metamorphic rocks that this chart is discussing.

  → **The symbols are used very often on diagrams on regents questions on other topics – you must know to look at this chart to identify which rock is which.**

- The chart is split into 2 parts under the **TEXTURE** column on the *far left*:

  → **FOLIATED** – the *mineral crystals are layered* as they're rearranged – usually happens with **regional metamorphism** (see below for explanation). Depending on the size of the crystals, foliation is known as either:
  - ✓ MINERAL ALIGNMENT or
  - ✓ BANDING

  → **NONFOLIATED** – the mineral crystals are *not layered*; rock is made up of *interconnected mineral crystals*

THE REFERENCE TABLES UNEARTHED | A Clear & Simple Reference Tables Guide

- The **GRAIN SIZE** column describes grain size of the rock:
  - ➔ Grain size ranges from *fine* to *coarse*

- The **COMPOSITION** column describes the mineral composition of the rocks.
  - ➔ **Foliated rocks** are mostly composed of the same six minerals – *pyroxene, garnet, amphibole, feldspar, quartz & mica*
    - ✓ Ex: *Slate* is composed of just *mica*
    - ✓ Ex: *Gneiss* is composed of all 6 minerals
    - ✓ Ex: *Phyllite* and *some Schist* is composed of all six minerals *but pyroxene*
  - ➔ **Nonfoliated rocks** are each composed of specific minerals identified on the chart.

- The **TYPE OF METAMORPHISM** column identifies the type of metamorphism the metamorphic rock underwent as it formed.
  - ➔ **Regional metamorphism** – *increase in temperature & pressure of an area/region* (often caused by plate movement) causes rocks to metamorphose.
    - ✓ Notice the arrow going down in the regional metamorphism area. It's showing that the lower rocks (*gneiss, schist*) were metamorphosed with **more heat and pressure** than the higher rocks (*slate, phyllite*) were.
  - ➔ **Contact metamorphism** – *contact with **magma** of an intrusion or **lava** of an extrusion* causes rocks to metamorphose.

- The **COMMENTS** column describes different properties specific to each kind of metamorphic rock not yet discussed on the table, such as what kind of rock the rock was before it metamorphosed.
  - ➔ Ex: **Slate** is a low-grade (not extreme) metamorphism of Shale
  - ➔ Ex: **Anthracite Coal** is a metamorphism of Bituminous Coal

→ If shown a diagram of sedimentary rock layers with an igneous intrusion and symbol of contact metamorphism, you must be able to identify the kind of rock that the sedimentary rock metamorphosed into when it encountered the intrusion. You can do this by:

- ✓ Identifying the sedimentary rock using the **MAP SYMBOL** column on the far right of the Scheme for Sedimentary Rock Identification table.
- ✓ Move over to this table - **Scheme for Metamorphic Rock Identification** and look in the **COMMENTS** column to find the metamorphic rock that formed from the rock you found on the other table.

Putting It All Together: Which metamorphic rock has a *foliated texture*, is *composed of mica, quartz, feldspar, amphibole, garnet and pyroxene* and has a *medium to coarse grain size*?

- "foliated texture" → slate, phyllite, schist, gneiss
- "composed of mica, quartz, feldspar, amphibole, garnet and pyroxene" → ~~slate~~, ~~phyllite~~, schist, gneiss
- "medium to coarse grain size" → ~~schist~~, **gneiss**

Base your answers to questions 50 and 51 on the flowchart below and on your knowledge of Earth science. The boxes labeled A through G represent rocks and rock materials. Arrows represent the processes of the rock cycle.
[See also RT 6 - Rock Cycle in Earth's Crust besides RT 7]

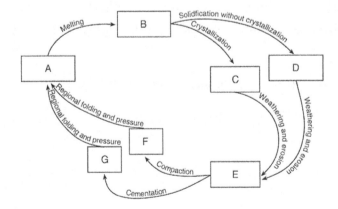

49. Which lettered box could represent the rock conglomerate?
(1) E   (3) C
(2) G   (4) D

50. The arrows in the block diagram below represent forces forming mountains in a region of Earth's lithosphere.

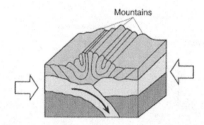

Metamorphic rocks that formed from these forces are represented by which lettered box in the flowchart?
(1) A
(2) B
(3) E
(4) F

51. Base your answer to this question on the geologic cross section below and on your knowledge of Earth science.

The geologic cross section represents rock layers of a portion of the Niagara Escarpment, and landscape features that are found in the Niagara region. The rock layers have not been overturned.

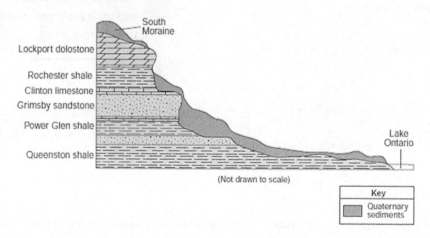

Identify the mineral composition of the Lockport dolostone. _____

52. One difference between a breccia rock and a conglomerate rock is that the particles in a breccia rock are
(1) more aligned
(2) more angular
(3) harder
(4) land derived

53. Base your answer to this question on the three bedrock outcrops below and on your knowledge of Earth science. The outcrops, labeled I, II, and III, are located within 15 kilometers of each other. Lines AB and CD represent unconformities. Line XY represents a fault. No overturning of the layers has occurred.

Which processes produced the brown siltstone layer in outcrops I and II?
(1) cooling and solidification of mafic lava at Earth's surface
(2) cooling and solidification of felsic magma deep within Earth
(3) compaction and cementation of rock fragments ranging in size from 0.006 to 0.2 centimeter in diameter
(4) compaction and cementation of rock fragments ranging in size from 0.0004 to 0.006 centimeter in diameter

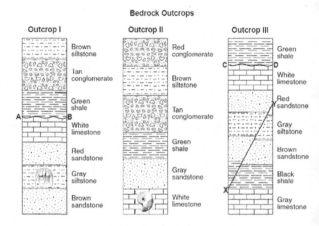

54. Base your answer to this question on the cross section below and on your knowledge of Earth science. On the cross section, numbers 1 through 7 represent rock units in which overturning has not occurred. Line XY represents a fault and line WZ represents the location of an unconformity.

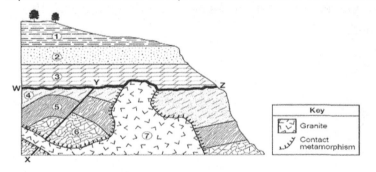

Name one sedimentary rock that was most likely metamorphosed to form rock unit 6.

_____

55. Which rock would be the best source of the mineral garnet?
   (1) basalt   (3) schist
   (2) limestone   (4) slate

56. What are the rock name and map symbol used to represent the sedimentary rock that has a grain size of 0.006 to 0.2 centimeters?

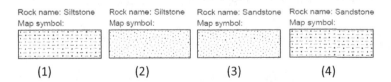

57. Which rock is composed of the mineral halite that formed when seawater evaporated?
   (1) limestone   (3) rock gypsum
   (2) dolostone   (4) rock sal

NEW YORK STATE
EARTH SCIENCE REFERENCE TABLES

PAGE 10

- Inferred Properties of Earth's Interior

Earth Science Reference Tables Page 10

Inferred Properties of Earth's Interior

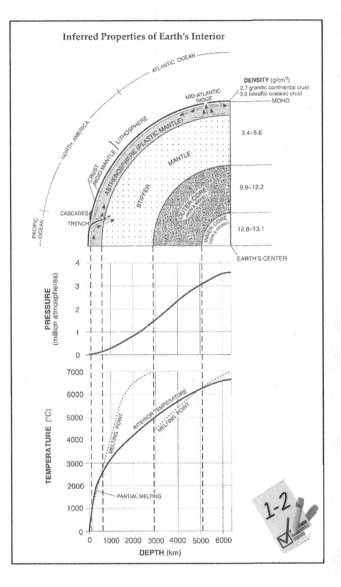

This table is a **cross-section** (side view) of *Earth's lithosphere* and *interior*.

The Table is Split in Two:

- The *top* shows **Earth's layers** and their **compositions, phases & densities**.

- The *bottom* shows the **temperature & pressure** of *each layer* as its **depth** increases.

Reading the Top Half of the Table:

Layers of Earth:

- ***Lithosphere:*** layer of Earth beneath the hydrosphere (oceans) & atmosphere. Composed of:

    → **CRUST**

      ✓ **Continental (land) crust** – granite, less dense; thicker
      ✓ **Oceanic crust** – basalt, denser; thinner

    → **RIGID MANTLE** – much denser than the crust

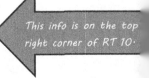

This info is on the top right corner of RT 10.

Earth Science Reference Tables Page 10

- **Earth's Interior:**
  - → **MANTLE:**
    - ✓ *Asthenosphere/plastic mantle* – partially molten (liquid)
    - ✓ Stiffer mantle – solid
  - → **OUTER CORE** – liquid, composed of iron & nickel
  - → **INNER CORE** – solid, composed of iron & nickel

> *The Asthenosphere...*
> ⇨ Detaches the lithosphere from the layers below and allows it to move independently
> ⇨ Is the source of magma (liquid molten rock)
> ⇨ Distributes its heat via convection currents (represented by **dark** arrows on the RT)

Don't memorize which layers are liquid & which are solid – just give a quick peek at the *bottom half* of the table:

If the INTERIOR TEMPERATURE line (solid dark line) is *above* the melting point line (dotted line), that means that the temperature is *hotter* than its melting point temperature, so it melted = liquid.

If the INTERIOR TEMPERATURE line is *below* the melting point line, that means the temperature is *cooler* than its melting point temperature, which means that it remains a solid.

Reading the Bottom Half of the Table
(and Density on Top Right):

- **DENSITY:** expressed in *g/cm³*
- **PRESSURE:** expressed in *millions of atmospheres*
- **TEMPERATURE:** expressed in *degrees Celsius*
- **DEPTH:** expressed in *kilometers*

As depth increases in Earth's interior, density, pressure, and temperature increase as well.

<u>Example</u>: *Characteristics of the **outer core***:

- → **Density:** ranges from *9.9* to *12.2 g/cm3*
- → **Pressure:** ranges from *1.5* to *3.1 million atm.*
- → **Temperature:** ranges from *5,000* to *6,300° C*
- → **Depth:** 2,900 to 5,100 km

58. The pressure at the interface between Earth's outer core and inner core is inferred to be
    (1) 0.2 million atmosphere
    (2) 1.5 million atmospheres
    (3) 3.1 million atmospheres
    (4) 3.6 million atmospheres

59. The convection currents responsible for moving tectonic plates occur in which Earth layer?
    (1) crust
    (2) rigid mantle
    (3) stiffer mantle
    (4) asthenosphere

60. In which Earth layer does the pressure reach 3.5 million atmospheres?
    (1) crust
    (2) stiffer mantle
    (3) outer core
    (4) inner core

61. The diagram below represents a model of Earth's surface and internal structure. Letters A, B, C, and D represent four different layers. Some depths below Earth's surface are shown.

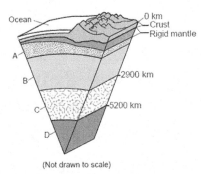

Which Earth layer is inferred to be composed of solid nickel and iron?
(1) A
(2) B
(3) C
(4) D

62. Base your answer to this question on the cross sections below and on your knowledge of Earth science. The cross sections represent three different stages in the development of Denali (Mt. McKinley) and the growth of the North American Plate in Alaska near the boundary with the Pacific Plate. Arrows represent the direction of plate movement.

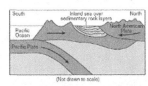
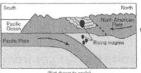
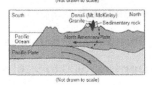

State the average density of the continental crust of the North American Plate and the average density of the oceanic crust of the Pacific Plate.

North American Plate continental crust:
_____ g/cm³
Pacific Plate oceanic crust:
_____ g/cm

NEW YORK STATE
EARTH SCIENCE REFERENCE TABLES

PAGE 11

- Earthquake P-Wave and S-Wave Travel Time

Earth Science Reference Tables Page 11

Earthquake P-Wave and S-Wave Travel Time

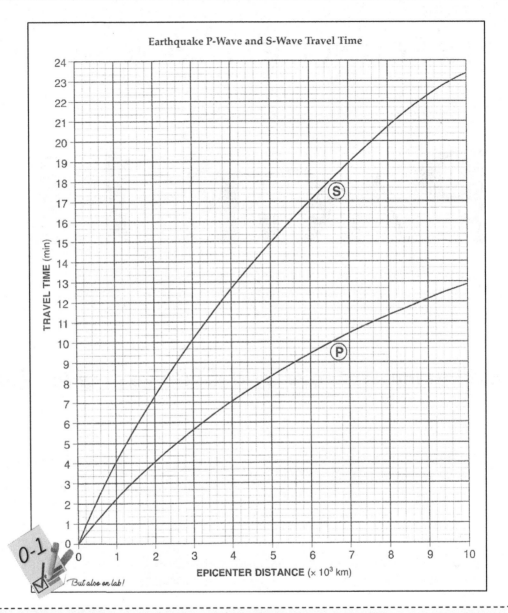

This table helps you figure out:

- The **epicenter distance** from a given **seismic station** (place where seismic waves are recorded on a seismograph)
- The **arrival time** of the **p-wave** or the **s-wave**
- The **starting time** of an **earthquake**

Earth Science Reference Tables Page 11

Reading the Table:

- The **X-axis** is **EPICENTER DISTANCE**, expressed in *thousands of kilometers* in scientific notation. Each **bold** line is **1,000 kilometers** and each *non-bold* line is **200 kilometers**.
  - → **Ex:** = 2,400 km
- The **Y-axis** is **TRAVEL TIME**, expressed in *minutes*. Each **bold** line is **1 minute**, and each *non-bold* line is **20 seconds**.
  - → **Ex:** = 3 min and 40 sec

To Figure Out...

Time Difference, Given *Epicenter Distance*:

- Go to the given **EPICENTER DISTANCE** on the *bottom*.
- Slide your finger *up* to the **P AND S-WAVE CURVES**
- Measure the distance between the two curves using a scrap paper
- Move the scrap paper to **TRAVEL TIME** on the *left* and **measure the time difference** between the two.

Example

Seismic station is 4,600 km from epicenter. P-wave arrived at 4:22 p.m. *When did the S-wave arrive?*

- → Go to 4,600 on EPICENTER DISTANCE axis
- → Slide finger *up* to P and S-wave curves
- → Measure the distance using a scrap paper
- → Move scrap paper over to TRAVEL TIME – see that time difference is approx. 6 min 20 sec
- → S-wave always arrives after P-wave. That means the S-wave arrived 6 min & 20 sec after 4:22 p.m. – at **4:28:20 p.m.**

The *further the* **distance** from the seismic station to the epicenter, the *larger the time difference* between the P & S-waves' arrival time.

Earth Science Reference Tables Page 11

Example

P-wave arrived at the seismic station at 12:02; S-wave arrived at 12:09. *How far is the epicenter?*

→ Time difference is 7 minutes.
→ Measure 7 minutes on the TRAVEL TIME axis with a scrap paper
→ Slide the paper between the P & S Wave curves until it fits perfectly
→ Follow the line *down* and see distance is **5,400 km** (5 & 2 lines X 10^3; each line is 200 km)

To Figure Out...

Epicenter Distance, Given *Time Difference*:

- Measure the time difference on **TRAVEL TIME** axis using a *scrap paper*
- Then **slide your paper** between the **P & S-WAVE CURVES** on the table until it fits perfectly
- Slide your finger *down* the line until you get the **EPICENTER DISTANCE**

Example

Seismic station is 4,600 km from epicenter. P-wave arrived at 4:22 p.m. *When did the earthquake begin?*

→ Go to 4,600 (4 & 3 lines) on EPICENTER DISTANCE
→ Slide finger *up* to P-wave curve
→ Slide finger *left* to TRAVEL TIME and see that P-wave arrived approximately 7 min and 50 sec after the earthquake started
→ Subtract that time from 4:22 and get **4:14:10 p.m.** – the time the earthquake began

To Figure Out...

Time Earthquake Began, Given *Epicenter Distance & P/S-Wave Arrival Time*:

- Go to your given **EPICENTER DISTANCE** on the *bottom*.
- Slide your finger *up* to the **P/S-WAVE CURVES** (whichever you're given)
- Slide your finger to the *left* to the **TRAVEL TIME** axis to see how many minutes after the earthquake began the P/S-wave arrived
- Subtract this time difference from your given time.
- The time you get is when the earthquake began.

To Figure Out...

Arrival Time of P/S-Wave, Given *Epicenter Distance & Arrival Time of the Other Wave:*

- Go to given **EPICENTER DISTANCE** on the *bottom*.
- Slide your finger *up* to the **P & S-WAVE CURVES**
- **Measure the distance** between the two curves using a scrap paper
- Move the scrap paper to the **TRAVEL TIME** axis on the *left* and **measure the time difference** between the two.
- **Add or subtract that time difference** to or from the time you are given to get your arrival time.
    → Since the P-wave always arrives before the S-wave:
    ✓ subtract from the S-wave
    ✓ add to the P-wave

Example

A seismic station is 3,800 km away from epicenter. The S-wave arrived at the station at 3:38 a.m. *When did P-wave arrive?*

→ The time difference is 5 minutes and 30 seconds (slide scrap paper between P & S waves on the 3,800-km line, then measure that distance on travel time bar)

→ P-waves arrives first – so P-wave arrived at **3:32:30 a.m.**

63. The epicenter of an earthquake was located 1800 kilometers from a seismic recording station. If the S-wave arrived at the seismic station at 10:06:40 a.m., at what time did the P-wave arrive at the same seismic station?
   (1) 10:03:00 a.m.
   (2) 10:03:40 a.m.
   (3) 10:09:40 a.m.
   (4) 10:10:20 a.m.

64. If a seismic station is 3200 km from an earthquake epicenter, what is the time needed for an S-wave to travel from the epicenter to the seismic station?
   (1) 4 min 40 sec
   (2) 6 min 0 sec
   (3) 10 min 40 sec
   (4) 11 min 10 sec

65. What is the approximate time difference between the first P-wave and the first S-wave recorded at a seismic station located 8000 kilometers from an earthquake's epicenter?
   (1) 8 minutes 40 seconds
   (2) 9 minutes 20 seconds
   (3) 11 minutes 20 seconds
   (4) 20 minutes 40 seconds

66. Base your answer to this question on the diagram below and on your knowledge of Earth science. The diagram represents a cut-away view of Earth's interior and the paths of some of the seismic waves produced by an earthquake that originated below Earth's surface. Points A, B, and C represent seismic stations on Earth's surface. Point D represents a location at the boundary between the core and the mantle.

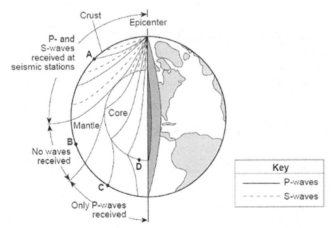

Seismic station A is 5000 kilometers from the epicenter. What is the difference between the arrival time of the first P-wave and the arrival time of the first S-wave recorded at this station?
   (1) 2 minutes 20 seconds
   (2) 6 minutes 40 seconds
   (3) 8 minutes 20 seconds
   (4) 15 minutes 00 second

NEW YORK STATE
EARTH SCIENCE REFERENCE TABLES

PAGE 12

- Dewpoint (°C)
- Relative Humidity (%)

Dewpoint (°C) / Relative Humidity (%)

Relative Humidity (%)

Dry-Bulb Temperature (°C)	Difference Between Wet-Bulb and Dry-Bulb Temperatures (C°)															
	0	1	2	3	4	5	6	7	8	9	10	11	12	13	14	15
-20	100	28														
-18	100	40														
-16	100	48														
-14	100	55	11													
-12	100	61	23													
-10	100	66	33													
-8	100	71	41	13												
-6	100	73	48	20												
-4	100	77	54	32	11											
-2	100	79	58	37	20	1										
0	100	81	63	45	28	11										
2	100	83	67	51	36	20	6									
4	100	85	70	56	42	27	14									
6	100	86	72	59	46	35	22	10								
8	100	87	74	62	51	39	28	17	6							
10	100	88	76	65	54	43	33	24	13	4						
12	100	88	78	67	57	48	38	28	19	10	2					
14	100	89	79	69	60	50	41	33	25	16	8	1				
16	100	90	80	71	62	54	45	37	29	21	14	7	1			
18	100	91	81	72	64	56	48	40	33	26	19	12	6			
20	100	91	82	74	66	58	51	44	36	30	23	17	11	5		
22	100	92	83	75	68	60	53	46	40	33	27	21	15	10	4	
24	100	92	84	76	69	62	55	49	42	36	30	25	20	14	9	4
26	100	92	85	77	70	64	57	51	45	39	34	28	23	18	13	9
28	100	93	86	78	71	65	59	53	47	42	36	31	26	21	17	12
30	100	93	86	79	72	66	61	55	49	44	39	34	29	25	20	16

Dewpoint (°C)

Dry-Bulb Temperature (°C)	Difference Between Wet-Bulb and Dry-Bulb Temperatures (C°)															
	0	1	2	3	4	5	6	7	8	9	10	11	12	13	14	15
-20	-20	-33														
-18	-18	-28														
-16	-16	-24														
-14	-14	-21	-36													
-12	-12	-18	-28													
-10	-10	-14	-22													
-8	-8	-12	-18	-29												
-6	-6	-10	-14	-22												
-4	-4	-7	-12	-17	-29											
-2	-2	-5	-8	-13	-20											
0	0	-3	-6	-9	-15	-24										
2	2	-1	-3	-6	-11	-17										
4	4	1	-1	-4	-7	-11	-19									
6	6	4	1	-1	-4	-7	-13	-21								
8	8	6	3	1	-2	-5	-9	-14								
10	10	8	6	4	1	-2	-5	-9	-14	-28						
12	12	10	8	6	4	1	-2	-5	-9	-16						
14	14	12	11	9	6	4	1	-2	-5	-10	-17					
16	16	14	13	11	9	7	4	1	-1	-6	-10	-17				
18	18	16	15	13	11	9	7	4	2	-2	-5	-10	-19			
20	20	19	17	15	14	12	10	7	4	2	-2	-5	-10	-19		
22	22	21	19	17	16	14	12	10	8	5	3	-1	-5	-10	-19	
24	24	23	21	20	18	16	14	12	10	8	6	2	-1	-5	-10	-18
26	26	25	23	22	20	18	17	15	13	11	9	6	3	0	-4	-9
28	28	27	25	24	22	21	19	17	16	14	11	9	7	4	1	-3
30	30	29	27	26	24	23	21	19	18	16	14	12	10	8	5	1

These Two Tables Help You Determine:

- The **DEWPOINT (°C)** or **RELATIVE HUMIDITY (%)**, given the *dry bulb temperature* and the *difference* between the dry bulb and wet bulb temperatures

- The **DRY/WET BULB TEMPERATURE**, given the *difference* between the two temperatures and the *dewpoint* or *relative humidity*

- *Relative humidity* – the amount of **water vapor** in the air *relative* to the amount the air can hold, expressed in a ***percentage***
  - **Ex:** If air can hold **10** grams of water vapor and now is holding **7** grams, the relative humidity is **70%**

- *Dewpoint* – the temperature when the **relative humidity is 100%** - the air is completely saturated (full of water vapor)

- *Psychrometer* – the *instrument* used to measure **relative humidity**. It's made of two thermometers:
  - A *wet bulb thermometer* – a thermometer wrapped in a wet cloth
  - A *dry bulb thermometer* – a regular thermometer measuring air temperature

Reading the Tables:

To Figure Out...

Both the Dewpoint and Relative Humidity Tables work the same way – just ensure you're using the correct table.

The Dewpoint/Relative Humidity, Given *Dry Bulb Temperature & Difference*

- Place one finger on the **DRY BULB TEMPERATURE** column on the *left* at the temperature you were given

- Place another finger on the **DIFFERENCE BETWEEN WET-BULB & DRY BULB TEMPERATURES** row on *top*

- Slide your two fingers *down* and *across* until they meet. This is the ***Dewpoint Temperature*** or ***Relative Humidity*** (depending on which chart you're using)

    → See <u>Example 1</u> below.

- Sometimes you are only given the two temperatures and you must figure out the **difference** yourself. To do that, *subtract the smaller number from the greater number* (greater minus smaller!) – the difference is always positive.

    → See <u>Example 2</u> below.

---

1. The dry bulb temperature is 12° C, and the difference between the wet and dry bulb temperatures is 4° C. What's the *relative humidity*?

    → Place your one finger on the dry bulb temperature on the left another finger on the difference on the *top*

    → Slide your fingers *down* & *across* and see that they meet at **57%**. That's the **relative humidity**.

2. The dry bulb temperature is 6° C and the wet bulb temperature is 5° C. What's the *relative humidity*? (You should know it's ^high because there is a small difference between the two numbers.)

    → Go to 6 on the DRY-BULB column.

    → Then **DO NOT** go to 5 on DIFFERENCE row. 5 is not the difference – it's the wet-bulb temperature. YOU MUST FIGURE OUT DIFFERENCE BETWEEN THE TWO!

    → Difference: 6 – 5 = **1° C**

    → Go to 1 on DIFFERENCE row. Slide fingers...

    → See that the relative humidity is **86%**.

Earth Science Reference Tables Page 12

To Figure Out...

The Dry/Wet Bulb Temperature, Given *Difference & Dewpoint/Relative Humidity*

- Place one finger on **DIFFERENCE** row or **DRY-BULB** column. (It depends on which information you're given.)

- Slide your finger *down/across* until you reach the **DEWPOINT** or **RELATIVE HUMIDITY** value you've been given

- Slide finger *across/up* until you hit the **DRY-BULB** column or **DIFFERENCE** row.

    → That's your answer, unless you're asked for the **wet-bulb temperature.**

    → When asked for the **wet-bulb temperature**, subtract the difference from the dry bulb temperature to get the wet bulb temperature. (Because the wet bulb is ALWAYS less than or equal to the dry bulb temperature.)

Example

The dry bulb temperature is 4° C. The dewpoint is -4° C. What is the *wet bulb temperature*?

→ Place your finger on 4 in DRY BULB column in **DEWPOINT** chart.

→ Slide across until you reach -4.

→ Slide up to the DIFFERENCE row. The difference is 3.

→ Since the *wet bulb temperature* is *always < the dry bulb temperature*, subtract 3 from 4 to get wet bulb temperature → 1° C.

To Figure Out...

Wet Bulb Temperature, Given *Relative Humidity/Dewpoint & Dry Bulb Temperature*

- Place your finger on given **DRY BULB TEMPERATURE** on the *left* column.

- Slide your finger *across* until you reach the given **RELATIVE HUMIDITY** or **DEWPOINT**.

- Slide your finger *up* until you reach the **DIFFERENCE BETWEEN WET-BULB & DRY-BULB TEMPERATURES** row.

- Take that **difference** and *subtract it* from given **dry-bulb temperature** (because wet-bulb temperature is always less than or equal to the dry bulb temperature).

More Practice:

Using both tables on page 12, fill out the following chart. See answers above.

Answers: 1) 10°C 2) 2°C 3) 16°C 4) 8°C 5) 5°C 6) 11% 7) 20°C 8) -2°C 9) 12°C 10) 6% 11) 16°C 12) 21%

Dry-Bulb Temperature	Wet-Bulb Temperature	Difference	Relative Humidity	Dewpoint
1)	2)	8° C	13%	-14° C
24° C	3)	4)	42%	10° C
0° C	-5° C	5)	6)	-24° C
7)	10° C	10° C	23%	8)
18° C	6° C	9)	10)	-19° C
11)	7° C	9° C	12)	-6° C

67. What is the dewpoint when the dry bulb temperature is 20°C and the relative humidity is 17%?
    (1) –5°C   (3) 11°C
    (2) –2°C   (4) 15°C

68. The dewpoint is 15°C. What is the wet-bulb temperature on a sling psychrometer if the dry-bulb temperature is 18°C?
    (1) 16°C   (3) 3°C
    (2) 2°C   (4) 20°C

69. If air has a dry-bulb temperature of 2°C and a wet-bulb temperature of –2°C, what is the relative humidity?
    (1) 11%   (3) 36%
    (2) 20%   (4) 67%

70. Base your answer to this question on the diagram below and on your knowledge of Earth science. The diagram represents a weather balloon as it rises from Earth's surface to 1000 meters (m). The air temperature and wet-bulb temperature values in degrees Celsius (°C) and the air pressure values in millibars (mb) are given for three altitudes.

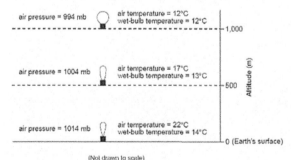

Determine the dewpoint and the relative humidity of the air at Earth's surface.
Dewpoint: ___ °C   **Relative humidity: ___ %**

New York State Earth Science Reference Tables

Page 13

- Temperature
- Pressure
- Key to Weather Map Symbols

Earth Science Reference Tables Page 13

Temperature

This table converts between the three **temperature** scales. **Fahrenheit**, **Celsius** and **Kelvin**.

Reading the Table:

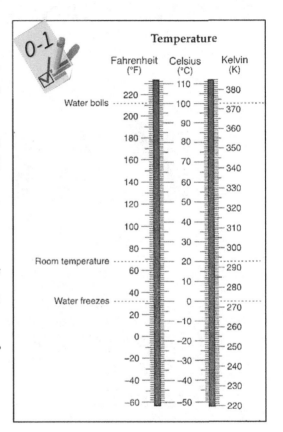

- To *convert* from one **temperature scale** to another: **slide your finger** or, for more accurate results, a **paper**, across the chart.

- Every small black line represents **2 degrees** in **Fahrenheit** and **1 degree** in **Celsius** and **Kelvin**.

    → **Ex:** Convert 170° F to °C and °K.
       ✓ 77° C, 350° K.

- Note the **boiling** and **freezing points** of **water** and **room temperature** indicated on the table.

    → **Ex:** Freezing point of water = **32° F/0° C/273° K**

 71. Base your answer to this question on the passage below and on your knowledge of Earth science.

Cosmic Microwave Background Radiation
In the 1960s, satellite probes found that cosmic microwave background radiation fills the universe uniformly in every direction and indicated a temperature of about 300 kelvins (K).

The current temperature indicated by the cosmic microwave background radiation is
(1) higher than the temperature at which water boils
(2) between the temperature at which water boils and room temperature
(3) between room temperature and the temperature at which water freezes
(4) lower than the temperature at which water freezes

Earth Science Reference Table Page 13

Pressure

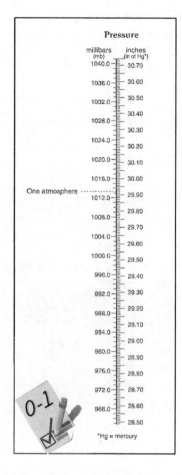

Atmospheric/Barometric/Air Pressure – the *weight of the air* pushing down onto Earth.

It's measured both in ***millibars***, with an ***aneroid barometer***, and with ***inches of mercury***, with a ***mercury* barometer**.

You must recognize this picture as a tool that measures air pressure.

ANEROID BAROMETER

The table **converts air/barometric pressure** from ***millibars (mb)*** to ***inches of mercury (in of Hg)*** and vice versa.

Reading the Table:

- Each **dark black line** on the millibar side (*left* side) is **1 millibar**, and each dark line on the inches side (*right* side) is **0.01** (one hundredth) **inches**.

- To ***convert*** between scales, slide **finger** or **paper** across the chart.
    - **Ex:** 1021.0 mb = 30.15 in of Hg
    - **Ex:** 984.5 mb ≈ 29.07 in of Hg

- The table also shows ***standard air pressure*** (air pressure at sea level) expressed in **atmospheres**.
    - Standard air pressure is **1 atm = 29.92 in of Hg = 1013.2 mb**

Earth Science Reference Tables Page 13

72. What is the average air pressure exerted by Earth's atmosphere at sea level, expressed in millibars and inches of mercury?
    (1) 1013.25 mb and 29.92 in of Hg
    (2) 29.92 mb and 1013.25 in of Hg
    (3) 1012.65 mb and 29.91 in of Hg
    (4) 29.91 mb and 1012.65 in of Hg

73. *The diagram below shows an instrument used in weather forecasting.

This instrument measures atmospheric
(1) wind speed
(2) wind direction
(3) pressure
(4) temperature

74. Base your answer to this question on the weather map below and on your knowledge of Earth science. The weather map shows atmospheric pressures, recorded in millibars (mb), at locations around a low-pressure center (L) in the eastern United States. Isobars indicate air pressures in the western portion of the mapped area. Point A represents a location on Earth's surface.

Convert the air pressure at location A from millibars to inches of mercury. _____ **in of Hg**

75. Base your answer to this question on the diagram below and on your knowledge of Earth science. The diagram represents a weather balloon as it rises from Earth's surface to 1000 meters (m). The air temperature and wet-bulb temperature values in degrees Celsius (°C) and the air pressure values in millibars (mb) are given for three altitudes.

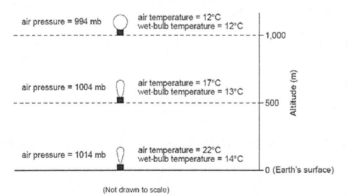

*Identify the names of the instruments carried by the weather balloon that recorded the air pressure and air temperature.

Air pressure: _____
Air temperature: _____

Key to Weather Map Symbols

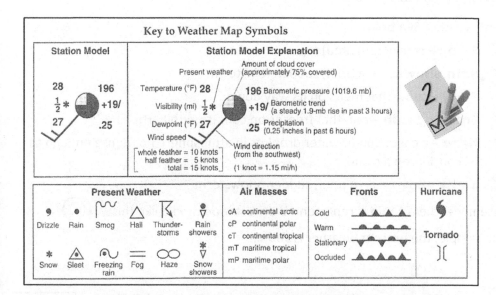

Station Model & Station Model Explanation

These two tables go hand in hand. The **station model** on the *right* explains the symbols on the station model on the *left*.

STATION MODEL – model used by meteorologists to represent **weather conditions**.

Reading the Table:

Starting from the *top left* and *counterclockwise* around the model:

- ***Temperature*** - measured with a **thermometer**; expressed in **°F**.
    - → An air temperature of **34° F** is represented like this on a station model:
        - ✓ **NOTE:** DO NOT WRITE "° F" on your model. The correct station model format for temperature is a number without the degree symbol or "F".
    - → On the sample, the temperature is **28° F**.

- ***Visibility*** – how far ahead you can see – measured in **miles**.
    - → On a station model, a visibility of 2½ miles is represented like this: 2½
    - → On the sample, visibility is **½ mile**.

Earth Science Reference Tables Page 13

- ***Present weather*** – describes the weather using the **PRESENT WEATHER symbols** on the *bottom left* of the table.
    → Present Weather Symbols:
    - ✓ **Smog** – smoky (polluted) fog
    - ✓ **Rain showers** – a brief, sudden rainstorm
    - ✓ **Sleet** – snow that melts on its way down
    - ✓ **Freezing rain** – rain that freezes on contact with Earth
    - ✓ **Haze** – a collection of water droplets in atmosphere not dense enough to be considered a cloud
    - ✓ **Snow showers** – a brief, sudden snowstorm
    → A present weather of **rain** is represented on a station model like this: •◯
    → On the sample, the present weather is *snow*.

- ***Dewpoint*** - measured with a ***psychrometer***; expressed in **°F**. Since the dewpoint table on RT 12 is expressed in degrees Celsius, be sure to *convert* to degrees Fahrenheit if using these two tables together.

    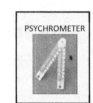

    → A dewpoint of 82° F is represented on a station model like this: ₈₂◯
    - ✓ DO NOT WRITE "° F" on the station model.
    → On the sample, the dewpoint is ***27° F***.

- ***Wind speed*** – measured with an ***anemometer***; measured in **knots** on the table, but often given in **miles per hour (mi/h)** as well.

    → *Conversion* from **knots** to **mi/h**: **1 knot = 1.15 mi/h** (on the table)
    - ✓ **Knots X 1.15 = mi/h** → 3 knots = 3.45 mi/h
    - ✓ **Mi/h ÷ 1.15 = knots** → 10 mi/h = 6.6 knots
    → Wind speed in knots is shown with **feathers** (lines) branching out of one main line:
    - ✓ A *whole* feather is **10** knots; a *half* feather is **5** knots
    - ✓ A wind speed of **35 knots** is represented like this on a station model:
    → On the sample, the wind speed is ***15 knots*** and ***17.25 mi/h***

Earth Science Reference Tables Page 13

- ***Wind direction*** – the direction the wind is **coming from**; measured with a ***wind vane***. The direction the stick is pointing is the direction the wind is coming from.

    → A wind direction of **northwest** is represented on station model like this:
    → On the sample, the wind is *coming from* the **southwest** (and *heading towards* the **northeast**).

- ***Precipitation*** – expressed in **inches in past 6 hours**; measured with a ***rain gauge/ruler***.

    → **0.68 inches of precipitation in the past 6 hours** is represented like this on a station model: ◯ .68
        ✓ **NOTE:** DO NOT WRITE "**0.68 in**" on your model. The correct station model format is without a zero before the decimal point – just "**.68**".
    → On the sample, **0.25 inches** of snow fell **in the past 6 hours**.

- ***Barometric Trend*** – the air pressure pattern – tells you whether the air pressure (barometric pressure) has steadily risen **(+/)** or steadily dropped **(-\)** in the **past 3 hours** and by how much.

    → To *convert* from **station model code** to **millibars**, *add* a decimal point between the two numbers.
        ✓ **Ex:** *+ 31/* → steady **3.1-mb rise** in the past 3 hours
        ✓ **Ex:** *- 22* → steady **2.2-mb drop** in the past 3 hours

    → To *convert* from **millibars** to **station model code**, *remove* the decimal between the two numbers.
        ✓ **Ex:** A steady *4.5-mb rise* in the past 3 hours → **+ 45/**
        ✓ **Ex:** A steady *1.7-mb drop* in the past 3 hours → **- 17**

    → On the sample, the barometric trend is a **steady 1.9-mb rise in the past 3 hours**.

- ***Barometric Pressure*** - expressed in a **3-digit station model code**. You must know how to *convert* this **code** into **millibars** and vice versa.

Earth Science Reference Tables Page 13

→ **Station Model Code → Millibars:**
- ✓ If the number is **less than 500**, place a **10** in front of the first digit
- ✓ If the number is **500 or greater**, place a **9** in front of the first digit.
- ✓ Add a **decimal point** before the last digit.

> 993 = **999.3 mb**
> 365 = **103**6.5 **mb**

→ **Millibars → Station Model Code:**
- ✓ Remove decimal
- ✓ Drop 10 or 9 from beginning of number

> **9**88.2 mb = 882
> **10**28.0 mb = 280

→ On the sample, the barometric pressure is **1019.6 mb**

Air Masses

Air Mass – a large **body of air** in the troposphere with similar **temperature**, **pressure** and **moisture** characteristics.

This table gives you the **symbols** of common **AIR MASSES** and what they stand for, but NOT any explanations or examples of source regions – these you must memorize. (See below.)

Air Masses:

- Are *named* for their *moisture content* and *temperature*
    - → **continental** (land) = dry
    - → **maritime** (water) = wet
    - → **arctic** = very cold
    - → **polar** = cold
    - → **tropical** = warm
- Acquire their unique properties from their *source region*
    - → **cA** = continental arctic → **North of Canada**
    - → **cP** = continental polar → **Central Canada**
    - → **cT** = continental tropical → **Southwest U.S.**
    - → **mT** = maritime tropical → **Gulf of Mexico**
    - → **mP** = maritime polar → **North Pacific, North Atlantic Oceans**

Air Masses	
cA	continental arctic
cP	continental polar
cT	continental tropical
mT	maritime tropical
mP	maritime polar

Tornado/Hurricane Symbols

This table shows you the **symbols** for **hurricanes** and **tornadoes**, the two most major kinds of **severe weather**. You must be able to identify these symbols on a weather map.

Hurricane – a severe tropical cyclone with *heavy rains* and *winds* moving at **75-150 mi/h**

Tornado – a violently destructive windstorm occurring over land characterized by a **funnel-shaped cloud** extending toward the ground

Earth Science Reference Tables Page 13

Fronts

> *Front* – **boundary** between two different air masses – a site of **rapid weather changes**
>
> This table gives you the **symbols** for the four kinds of fronts, but *no explanations*. You must understand what kind of air masses each kind of front is composed of and what kind of weather each front will cause.

Fronts:

- **Cold Front** – **cP** (cold) overtakes **mT** (warm):
    - → steep slope along boundary & warm air rises very quickly
    - → this causes a short period of heavy rain
    - → area is left with **cold** weather

- **Warm front** – **mT** (warm) advances towards **cP** (cold):
    - → warm air mass slowly rises over the cold air mass
    - → this causes long periods of gentle rain
    - → area is left with **warm** weather

- **Occluded front** – **cold front** overtakes a **warm front**:
    - → very **stormy**, unstable weather

- **Stationary front** – **mT** is near **cP**:
    - → neither is strong enough to push the other away
    - → there's precipitation along the boundary

Reading the Table:

The front symbol **points** *in the direction* the front is advancing.

Ex: The symbol to the left is a cold front coming from the west and traveling east.

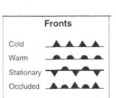

76. *Which weather instrument is used to measure air temperatures recorded on a weather map?
   (1) anemometer
   (2) wind vane
   (3) thermometer
   (4) barometer

77. *Which weather map symbol is associated with extremely low air pressure?

   (1) (2) (3) (4)

78. *Which area is the most common source region for cold, dry air masses that move over New York State?
   (1) North Atlantic Ocean
   (2) Gulf of Mexico
   (3) Central Canada
   (4) Central Mexico

79. Which weather map symbol is used to represent violently rotating winds that have the appearance of a funnel-shaped cloud?

   (1) (2) (3) (4)

80. The weather station model below shows some of the weather data for a certain location.

   What is the wind speed shown on the station model and *which instrument is used to measure the wind speed?
   (1) 15 knots, measured by a wind vane
   (2) 15 knots, measured by an anemometer
   (3) 25 knots, measured by a wind vane
   (4) 25 knots, measured by an anemometer

81. Which station model shows an air temperature of 75°F and a barometric pressure of 996.3 mb?

   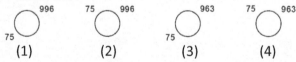

   (1) (2) (3) (4)

82. *Which station model represents a location that has the greatest chance of precipitation?

Earth Science Reference Tables Page 13

83. The table below lists some weather conditions for another location on this map.

Temperature (°F)	Dewpoint (°F)	Precipitation (inches in past 6 hours)	Present Weather
76	74	0.85	Rain showers

On the weather station model below, *using the proper format*, record the weather conditions listed in the table.

84. Base your answer to this question on the graphs below and on your knowledge of Earth science. The graphs show air temperatures and dewpoints in °F, and wind speeds in knots (kt) from 2:00 a.m. to 11:00 p.m. at a certain New York State location.

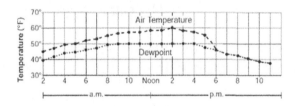

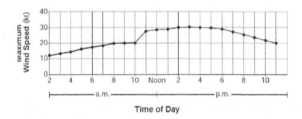

Time of Day

Which station model represents the weather data for this location at 4:00 p.m.?

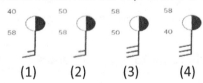

(1)   (2)   (3)   (4)

85. A weather station recorded the barometric pressure on a weather station model as shown below.

On the map below, place an X to represent a possible location for this weather station.

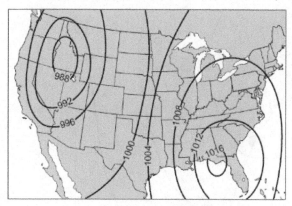

86. Base your answer to this question on the map below and on your knowledge of Earth science. The map shows surface air temperatures for some locations in the United States on a day in November. The 20°F, 30°F, 40°F, and 70°F isotherms are shown. Points A, W, X, Y, and Z represent locations on Earth's surface.

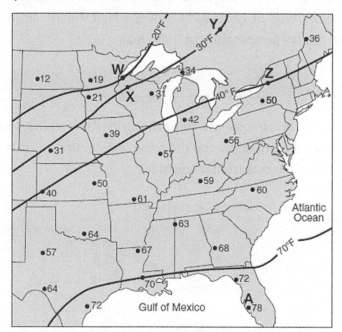

Identify the air temperature at Watertown, New York. ____°F

[See also Map of New York State on RT 2-3. This is a *very* challenging question!]

Earth Science Reference Tables Page 13

Base your answers to questions 88 and 89 on the weather map below and on your knowledge of Earth science.

The map indicates the location of a low-pressure system over New York State during late summer. Isobar values are recorded in millibars. Shading indicates regions receiving precipitation. The air masses are labeled mT and cP. The locations of some New York State cities are shown. Points A and B represent other locations on Earth's surface.

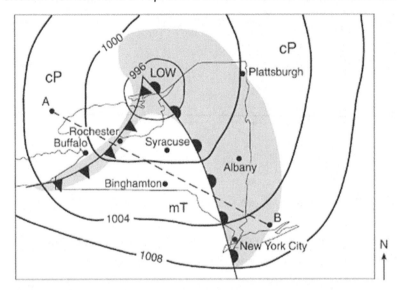

87. An air mass acquires the characteristics of the surface over which it forms. Circle either "land" or "ocean" below" to describe the type of Earth surface over which the mT air mass most likely formed and describe the relative temperature of that area.

    Circle one: **land   ocean**

    Relative temperature of Earth's surface: _____

88. The cross section below represents the atmosphere along the dashed line from A to B on the map. The warm frontal boundary is already shown on the cross section. Draw a curved line to represent the shape and location of the cold frontal boundary

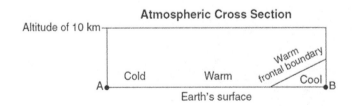

New York State Earth Science Reference Tables

Page 14

- Selected Properties of Earth's Atmosphere
- Planetary Wind and Moisture Belts
- Electromagnetic Spectrum

Earth Science Reference Tables Page 14

Selected Properties of Earth's Atmosphere

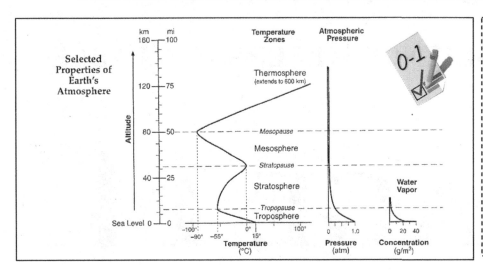

Atmosphere – the layer of air surrounding Earth.
This table describes the **ALTITUDE, TEMPERATURE, PRESSURE** and **WATER VAPOR CONCENTRATION** of each of the **atmosphere's layers**.

Reading the Table:

- The atmosphere is *subdivided* into **4 layers**, with a *"pause"* dividing each zone.

    → **Ex:** The *stratosphere* ranges from the **tropopause** to the **stratopause**.

- The column furthest to the *left* shows the **ALTITUDE** of the layers in both **kilometers (km)** and **miles (mi)**:

    → Every line on the *km* side is **10 km**
    → Every line on the *mi* side is **5 mi**

    Altitude of *Stratosphere* ranges from:
    - 12-50 km
    - 7-31 mi

- The **TEMPERATURE** area shows the temperature ranges of each of the atmosphere's zones in **°C**.

    → As altitude increases, temperature...
      ✓ *decreases* in Troposphere
      ✓ *increases* in Stratosphere
      ✓ *decreases* in Mesosphere
      ✓ *increases* in Thermosphere

    Temperature of *Stratosphere* ranges from -55° C to 0° C.

- **ATMOSPHERIC PRESSURE** shows the amount of *air pressure* in each layer measured in **atmospheres (atm)**.

→ As **altitude** increases, **pressure** decreases (because there's less air above...).

✓ **Ex:** Pressure in *Stratosphere* ranges from **0.25 atm to 0.1 atm**.

- **WATER VAPOR CONCENTRATION** shows the amount of water in gas form (humidity) in the atmosphere expressed in **grams per cubic meter (g/m3).**

    → Note that water vapor basically only exists in the **troposphere** – therefore, **weather** is limited to this atmospheric zone.

89. As altitude increases in the troposphere and stratosphere, the air temperature
    (1) decreases in the troposphere and increases in the stratosphere
    (2) decreases in both the troposphere and stratosphere
    (3) increases in the troposphere and decreases in the stratosphere
    (4) increases in both the troposphere and stratosphere

90. The ozone layer protects life on Earth by absorbing harmful ultraviolet radiation. The ozone layer is located between 17 kilometers and 35 kilometers above Earth's surface in which atmospheric temperature zone?
    (1) troposphere   (3) mesosphere
    (2) stratosphere   (4) thermosphere

91. Base your answer to this question on the passage below and on your knowledge of Earth science.

    Comets and Asteroids

    Since comets and asteroids orbit the Sun, both are part of our solar system. As they move through space, comets leave a debris trail of mostly dust-sized particles. When Earth passes through this debris, a meteor shower occurs, often filling the night sky with "shooting star" trails as they burn up in the atmosphere 50 to 80 kilometers above Earth's surface.

    In which temperature zone of Earth's atmosphere will most meteors burn up?
    (1) troposphere   (3) mesosphere
    (2) stratosphere   (4) thermosphere

92. Base your answer to this question on the data table to the right and on your knowledge of Earth science. The data table shows the average level of atmospheric carbon dioxide (CO_2), measured in parts per million (ppm), for the month of February at the Mauna Loa observatory in Hawaii from 2008 to 2014.

    These measurements of atmospheric carbon dioxide were collected at an altitude of 3.4 kilometers. Identify the temperature zone of the atmosphere where these data were collected. _____

Year	Average February Atmospheric CO_2 Levels (ppm)
2008	386
2009	387
2010	390
2011	392
2012	394
2013	396
2014	398

Earth Science Reference Tables Page 14

Planetary Wind & Moisture Belts in the Troposphere

Planetary/Prevailing Winds – strong winds that blow from one latitude to another in a specific direction. These winds form because of **pressure differences** at each latitude.

Planetary Wind Belts – the areas on Earth where the planetary winds blow.

Ex: The area from **0° to 30° N** is a **PLANETARY WIND BELT**. It has **northeasterly winds**, winds that blow from northeast to southwest.

Moisture Belts – areas that are categorized by extremely wet or extremely dry climates.

Ex: 0° - the **Equator** – is a wet belt. It has a *rainy* climate.

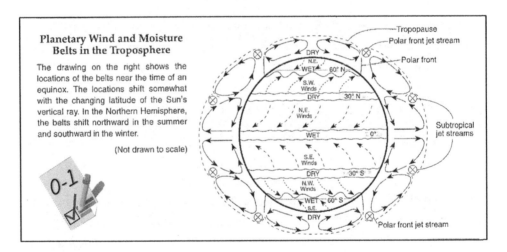

Reading the Table:

- The *curved dashed arrows* show the **direction** of **prevailing winds** within each **wind belt**. The winds alternate between **SW** & **SE** and **NW** & **NE**, depending on their origin.

  → The arrows are *curved* because the winds are **deflected** by the *Coriolis Effect*...

    ✓ To the *right, clockwise* in the **N hemisphere**

    ✓ To the *left, counterclockwise* in the **S hemisphere**

Earth Science Reference Tables Page 14

✓ Winds are deflected *from direction of origin*! (If the winds are coming from the North Pole, they are deflected to the right, but if you're looking at the globe head on, they seem to be deflected to the left.)

- Air always *moves* from areas of **high pressure** to areas of **low pressure**.
  - → 30° **N & S** and 90° **N & S** must be areas of **high pressure**, since winds are *diverging* (moving away) from those areas
  - → 0° **latitude** and 60° **N & S** must be areas of **low pressure**, since winds are *converging* (coming together) in those areas.
    - ✓ **HINT:** "Dry" rhymes with "High." "Dry" areas have **high** pressure.

- The **dark arrows** outside the circle represent **convection currents** in the atmosphere. The *bottom part* of each current is the **wind**.
  - → **Wind** = *horizontal movement of air* parallel to Earth's surface.

- The *dotted line* around the chart is the **tropopause**, which marks the end of the **troposphere**. This shows there are convection currents only in the troposphere.
  - **Ex:** At 0°, warm air rises ["vertical" lines are going straight "up" into troposphere]. Then cool air lands by 30° and completes the convection cell – goes back to 0° where it started. The bottom part of the convection current is the prevailing wind. [Notice the wind's direction. It goes from 30° to 0° and from 30° to 60°…]

- **Jet Streams** – bands of **fast moving air** (over 200 mph!) in the *troposphere* that blow from W-E. Represented by this symbol: ⊗

- **NOTE:** This table is based on the **equinoxes**, when the sun's vertical ray is on the **equator** and thus causes it to have the **lowest pressure** (because hot air has low pressure). The wind belts shift north in the summer, and south in the winter. (See paragraph to the *left* of the table.

- *Storms* usually travel from **SW** → **NE** (a normal "storm track") in the United States, because the U.S. is in a **southwesterly** planetary wind belt

Copyright 2019 © Y. Finkel | ALL RIGHTS RESERVED

93. The seasonal shifts of Earth's planetary wind and moisture belts are due to changes in the
(1) distance between Earth and the Sun
(2) amount of energy given off by the Sun
(3) latitude that receives the Sun's vertical rays
(4) rate of Earth's rotation on its axis

94. Near which two latitudes are most of Earth's dry climate regions found?
   (1) 0° and 60° N
   (2) 0° and 30° S
   (3) 30° N and 60° N
   (4) 30° N and 30° S

95. Jet stream winds over the United States generally move from
   (1) east to west
   (2) north to south
   (3) west to east
   (4) south to north

96. In which planetary wind belt do most storms move toward the northeast?
   (1) 30° N to 60°N
   (2) 0° to 30° N
   (3) 0° to 30° S
   (4) 30° S to 60° S

97. The map of North America below shows the position of the polar front jet stream on January 7, 2014, and the location of Atlanta, Georgia.

Which type of air mass was most likely located over Atlanta, Georgia? [See also RT 13 – Air Masses.]
(1) mT
(2) mP
(3) cT
(4) cP

98. The arrow on the map below represents the direction a wind is blowing over a land surface in the Northern Hemisphere without showing the Coriolis effect. Which dashed arrow represents how the wind direction will change in the Northern Hemisphere due to the Coriolis effect?

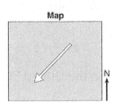

Base your answers to questions 100 and 101 on the weather map below and on your knowledge of Earth science. The weather map shows atmospheric pressures, recorded in millibars (mb), at locations around a low-pressure center (L) in the eastern United States. Isobars indicate air pressures in the western portion of the mapped area.

99. *Identify the weather instrument that was used to measure the air pressures recorded on the map.

_____

100. Identify the compass direction toward which the center of the low-pressure system will move if it follows a typical storm track. _____

101. Base your answer to this question on the passage and the map of South America below and on your knowledge of Earth science.

Two South American Deserts

South America is an excellent example of the influence that plate tectonic features have on climates. The Andes mountain range, formed by plate tectonics, is on the western edge of South America. When prevailing winds come from the southeast, which usually occurs between 0° and 30° S latitudes, rainfall is increased on the eastern side of the mountain range. The Atacama Desert lies in the rain shadow (dry area) to the west of the mountains. Farther south, the reverse pattern is found, due to different prevailing winds blowing between 30° S and 60° S latitudes. The Patagonian Desert lies on the eastern side of the Andes, between the Andes and the South Atlantic Ocean.

On the map below, draw one arrow in the box located on the Andes Mountains to indicate the surface planetary wind direction that helped produce the Atacama Desert.

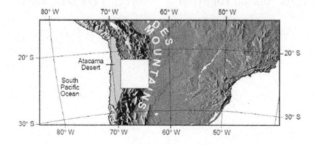

Electromagnetic Spectrum

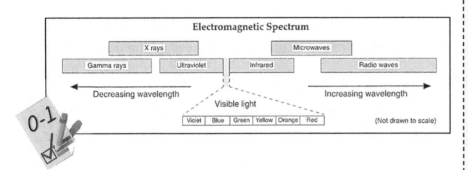

Reading the Table:

- **Gamma Rays** – **shortest** wavelength & **most** energized

- **Radio Waves** – **longest** wavelength & **least** energized

- **Visible Light** - the only EME that is visible. It is seen as sunlight – a white light – unless it is broken down in a prism into its 6 colors.
  - → **Red** is longer wavelength, less energized
  - → **Violet/blue** is shorter wavelength, more energized

- **Doppler's Effect** – an apparent **color/wavelength shift** as Earth and another celestial object move towards or away from each other.
  - → **Moving away:** wavelengths "stretch out" = *longer* = **RED shift**
  - → **Coming closer:** wavelengths "bunch together" = *shorter* = **BLUE shift**

Electromagnetic Energy (EME) – energy emitted by **moving particles** that travels in the form of **waves**

This table classifies electromagnetic energy by **wavelength**.

102. Which color of visible light has the shortest wavelength?
    (1) violet
    (2) green
    (3) yellow
    (4) red

103. Which diagram best represents the relative wavelengths of visible light, ultraviolet energy, and infrared energy?

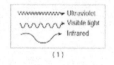
(1)

(2)

(3)

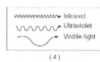
(4)

New York State Earth Science Reference Tables

Page 15

- Characteristics of Stars
- Solar System Data

Earth Science Reference Tables Page 15

Characteristics of Stars

Star - large ball of gas that produces energy and luminosity (shine) through **nuclear fusion**, a chemical process in which smaller atoms combine to form larger atoms.

This chart shows the **SIZE**, **COLOR**, **TEMPERATURE** & **LUMINOSITY** of stars, as well as their **GROUPING**.

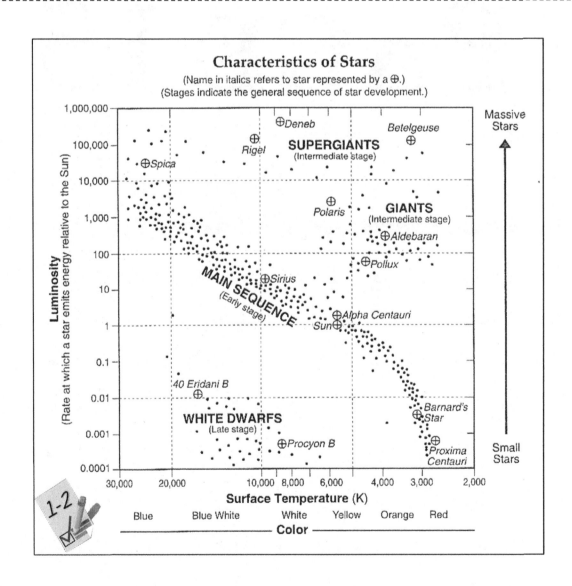

Reading the Table:

- **SIZE** – *right* side
  - → Stars range in *size* from **small** to **massive**
  - → The *sun* is a *medium-sized* star

- **SURFACE TEMPERATURE/COLOR** - *bottom*
  - → *Temperature* ranges from **2,000-30,000° K** (R →L gets increasingly hotter)
  - → *Colors* go in order of the **Electromagnetic Spectrum** (RT 14)
    - ✓ **Red** is the *coolest* and least energized
    - ✓ **Blue** is the *hottest* and most energized
  - → The *sun's* temperature is about *5,500° K* and the *sun's* color is *yellow-white*

- **LUMINOSITY** – *left* side
  - → *Luminosity* – brightness – depends on star's **size**; measured *relative to the sun*:
    - ✓ The *sun* has luminosity of **1**
    - ✓ *Brighter* than sun has a luminosity **greater than 1**
    - ✓ *Less bright* than the sun has a luminosity **less than 1**
  - → **Ex:** **Spica** is about *50,000 times more luminous* than the *sun*.

- **GROUPING:** Most stars fall into the *main sequence pattern*, which means that as *size* and *luminosity* increase, *temperature* increases as well.
  - → Some **exceptions** that don't follow this trend are the **white dwarfs** (*hot & small*) and **[red] giants** (*cool & large*).

104. Which star is more massive than our Sun, but has a lower surface temperature?
    (1) 40 Eridani
    (2) Sirius
    (3) Aldebaran
    (4) Barnard's Star

105. Which characteristics best describe the star Betelgeuse?
    (1) reddish orange with low luminosity and high surface temperature
    (2) reddish orange with high luminosity and low surface temperature
    (3) blue white with low luminosity and low surface temperature
    (4) blue white with high luminosity and high surface temperature

106. Which sequence of stars is listed in order of increasing luminosity?
    (1) Spica, Rigel, Deneb, Betelgeuse
    (2) Polaris, Deneb, 40 Eridani B, Proxima Centauri
    (3) Barnard's Star, Alpha Centauri, Rigel, Spica
    (4) Procyon B, Sun, Sirius, Betelgeuse

107. The diagram below represents a model of the size of the Sun and indicates the color of the Sun. Which diagram best represents the relative size and indicates the color of Polaris compared to the Sun?

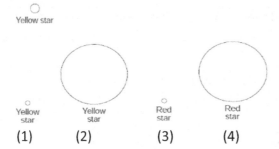

108. Compared to the Sun, the star Betelgeuse is
    (1) less luminous and warmer
    (2) less luminous and cooler
    (3) more luminous and warmer
    (4) more luminous and cooler

Earth Science Reference Tables Page 15

Solar System Data

This table lists ten **CELESTIAL OBJECTS** – Sun, 8 Planets and Earth's Moon, and gives various data about each one.

Solar System Data

Celestial Object	Mean Distance from Sun (million km)	Period of Revolution (d=days) (y=years)	Period of Rotation at Equator	Eccentricity of Orbit	Equatorial Diameter (km)	Mass (Earth = 1)	Density (g/cm³)
SUN	—	—	27 d	—	1,392,000	333,000.00	1.4
MERCURY	57.9	88 d	59 d	0.206	4,879	0.06	5.4
VENUS	108.2	224.7 d	243 d	0.007	12,104	0.82	5.2
EARTH	149.6	365.26 d	23 h 56 min 4 s	0.017	12,756	1.00	5.5
MARS	227.9	687 d	24 h 37 min 23 s	0.093	6,794	0.11	3.9
JUPITER	778.4	11.9 y	9 h 50 min 30 s	0.048	142,984	317.83	1.3
SATURN	1,426.7	29.5 y	10 h 14 min	0.054	120,536	95.16	0.7
URANUS	2,871.0	84.0 y	17 h 14 min	0.047	51,118	14.54	1.3
NEPTUNE	4,498.3	164.8 y	16 h	0.009	49,528	17.15	1.8
EARTH'S MOON	149.6 (0.386 from Earth)	27.3 d	27.3 d	0.055	3,476	0.01	3.3

Reading the Table:

- **CELESTIAL OBJECT**

  → The *celestial objects* listed in this column include the **sun**, the **eight planets**, and **Earth's moon**.

  ✓ Everything on the table is referring to these celestial objects.

  → *Planets* are divided into two major groups: **Terrestrial** (solid) planets and **Jovian** (gaseous) planets

  ✓ *Terrestrial* planets begin with **Mercury** and end with **Mars**.

  ✓ *Jovian* planets begin with **Jupiter** and end with **Neptune**

  → Using the table, you must be able to *compare* terrestrial and Jovian planets.

  Ex: List two differences between terrestrial and Jovian Planets.

  ✓ Terrestrial planets have higher densities and shorter periods of revolution than Jovian planets.

Earth Science Reference Tables Page 15

- **MEAN DISTANCE FROM SUN** – measured in **millions of kilometers**
    - "mean" – *average*. Because planets move along an **elliptical** (not perfectly circular) orbit, they do not constantly remain at the same distance from the sun.
    - Notice that the **terrestrial** planets (Mercury – Mars) are much *closer* to the sun than the **Jovian** planets are.
    - The *asteroid belt* (area where most asteroids orbit) is located **between Mars and Jupiter**. Therefore, most asteroids are found *between 227 and 778 million km* from the sun.

- **PERIOD OF REVOLUTION** – measured in **days (d)** and **years (y)**
    - *Period of revolution* – the amount of time it takes for a planet to make one full **revolution** around the **sun** → that planet's **year**
        - ✓ The **moon's** period of revolution – the amount of time it takes the moon to orbit *Earth*.
    - Earth's *period of revolution* is 365.26 days → 1 Earth year
        - ✓ Earth's *rate of revolution* is approximately **1°/day** (360° in 365 days)
    - The *closer* a planet is to the *sun*, the *shorter* its period of revolution and vice versa. Therefore…
        - ✓ *Terrestrial* planets have *shorter* periods of revolution
        - ✓ *Jovian* planets have *longer* periods of revolution

- **PERIOD OF ROTATION AT EQUATOR** – measured in **days (d), hours (h), minutes (min)** and **seconds (s)**
    - *Period of rotation* – the amount of time it takes a planet to make one full rotation/spin on its (imaginary) axis → that planet's **day**
        - ✓ **Ex:** The length of **Saturn's day** is *10 hours*
    - Earth's *period of rotation* is approximately **24 hours** → 1 [Earth] day.
        - ✓ Earth's *rate of rotation* is **15°/hour** (360° in 24 hours)

- **ECCENTRICITY OF ORBIT** – expressed in a **decimal** format

    → *Eccentricity* – how **oval/elliptical** an orbit is
    - ✓ The **greater** the number, the more eccentric/less circular it is. Mercury's orbit is *most eccentric*.
    - ✓ The **smaller** the number, the less eccentric/more circular it is. Venus's orbit is *least eccentric*.

- **EQUATORIAL DIAMETER** – measured in **kilometers (km)**

    → *Diameter* – size - how large the planet is
    - ✓ **Earth** & **Venus** have *similar diameters*
    - ✓ **Mercury** has the *smallest* diameter; **Jupiter** has the *largest* diameter

- **MASS** – as *compared to* **Earth**

    → *Mass* – the amount of matter in a substance
    → There's no unit – all planets' masses are **compared to Earth's mass**.
    - ✓ **Earth** – mass of 1
    - ✓ **Mars** = 0.06 of Earth's mass
    - ✓ **Sun** = 333,000 of Earth's mass

    > *Notice that the moon's period of rotation and period of revolution is the same – **27.3** days. This means it takes the same amount of time for the moon to rotate once around "itself" as it takes for the moon to revolve around Earth. **This explains why we always see the <u>same side of the moon</u>.***

- **DENSITY** – expressed in **grams per cubic centimeter (g/cm3)**

    → *Terrestrial* planets have **higher** densities because they are *solid*.
    → *Jovian* planets have **lower** densities because they are *gaseous*.
    - ✓ **Saturn** is the only planet that is less dense than water.

    (See **RT 1 – Properties of Water**.)

Earth Science Reference Tables Page 15

109. Which planet has a density that is less than the density of liquid water?
(1) Mercury (3) Mars
(2) Earth (4) Saturn

110. The asteroid Ceres lies at an average distance of 414 million kilometers from the Sun. The period of revolution of Ceres around the Sun is approximately
(1) 438 days (3) 4.6 years
(2) 687 days (4) 12.6 years

111. *Compared to terrestrial planets, Jovian planets have
(1) smaller equatorial diameters and shorter periods of revolution
(2) smaller equatorial diameters and longer periods of revolution
(3) larger equatorial diameters and shorter periods of revolution
(4) larger equatorial diameters and longer periods of revolution

112. *The diagram below represents the interiors of three planets in our solar system.

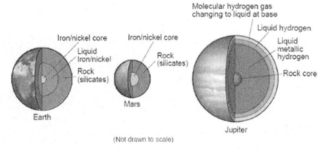

Which inference best describes the interiors of the planets in our solar system?
(1) Both terrestrial and Jovian planets have layered interiors, with density decreasing toward the center.
(2) Both terrestrial and Jovian planets have layered interiors, with density increasing toward the center.
(3) Only terrestrial planets have layered interiors, with density decreasing toward the center.
(4) Only Jovian planets have layered interiors, with density increasing toward the center.

113. Base your answer to this question on the passage below and on your knowledge of Earth science.

Comets and Asteroids

Since comets and asteroids orbit the Sun, both are part of our solar system. Asteroids are rocky objects that vary greatly in size. Most asteroids follow orbits between 300 and 600 million kilometers from the Sun, but several have been pulled from this region by the gravitational attraction of nearby planets.

Between which two planets are most asteroids located?
(1) Earth and Mars (3) Jupiter and Saturn
(2) Mars and Jupiter (4) Saturn and Uranus

114. *Explain how the Moon's rotation and revolution cause the same side of the Moon to always face Earth. _____

Base your answers to questions 117 through 118 on the diagram below and on your knowledge of Earth science. The diagram represents a planetary system, discovered in 2013, with seven exoplanets (planets that orbit a star other than our Sun) labeled b through h orbiting a star. The exoplanet orbits are represented with solid lines. For comparison, the orbits of three planets of our solar system are shown with dashed lines. The sizes of the star, exoplanets, and planets are not drawn to scale.

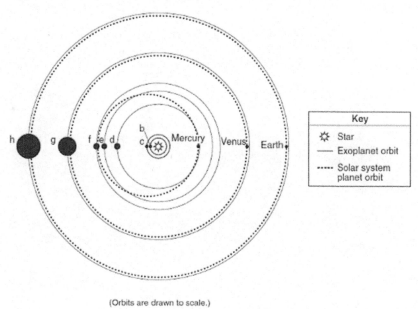

(Orbits are drawn to scale.)

115. Identify the name of the planet represented in the diagram that has the most eccentric orbit. _____

116. *Circle either "Terrestrial planet" or "Jovian planet" below to indicate the classification of the three solar system planets shown in the diagram. Describe one characteristic of this type of planet that distinguishes it from the other type of planet.
Circle one: **Terrestrial planet   Jovian planet**
Characteristic of this type of planet: _____

117. Identify the celestial object in our solar system that has a period of rotation that is most similar to the period of rotation of Earth's Moon. _____

NEW YORK STATE
EARTH SCIENCE REFERENCE TABLES

PAGE 16

- Properties of Common Minerals

Properties of Common Minerals

LUSTER	HARD-NESS	CLEAVAGE	FRACTURE	COMMON COLORS	DISTINGUISHING CHARACTERISTICS	USE(S)	COMPOSITION*	MINERAL NAME
Metallic luster	1–2		✔	silver to gray	black streak, greasy feel	pencil lead, lubricants	C	Graphite
Metallic luster	2.5		✔	metallic silver	gray-black streak, cubic cleavage, density = 7.6 g/cm³	ore of lead, batteries	PbS	Galena
Metallic luster	5.5–6.5		✔	black to silver	black streak, magnetic	ore of iron, steel	Fe_3O_4	Magnetite
Metallic luster	6.5		✔	brassy yellow	green-black streak, (fool's gold)	ore of sulfur	FeS_2	Pyrite
Either	5.5–6.5 or 1		✔	metallic silver or earthy red	red-brown streak	ore of iron, jewelry	Fe_2O_3	Hematite
Nonmetallic luster	1	✔		white to green	greasy feel	ceramics, paper	$Mg_3Si_4O_{10}(OH)_2$	Talc
Nonmetallic luster	2		✔	yellow to amber	white-yellow streak	sulfuric acid	S	Sulfur
Nonmetallic luster	2	✔		white to pink or gray	easily scratched by fingernail	plaster of paris, drywall	$CaSO_4 \cdot 2H_2O$	Selenite gypsum
Nonmetallic luster	2–2.5	✔		colorless to yellow	flexible in thin sheets	paint, roofing	$KAl_3Si_3O_{10}(OH)_2$	Muscovite mica
Nonmetallic luster	2.5	✔		colorless to white	cubic cleavage, salty taste	food additive, melts ice	$NaCl$	Halite
Nonmetallic luster	2.5–3	✔		black to dark brown	flexible in thin sheets	construction materials	$K(Mg,Fe)_3$ $AlSi_3O_{10}(OH)_2$	Biotite mica
Nonmetallic luster	3	✔		colorless or variable	bubbles with acid, rhombohedral cleavage	cement, lime	$CaCO_3$	Calcite
Nonmetallic luster	3.5	✔		colorless or variable	bubbles with acid when powdered	building stones	$CaMg(CO_3)_2$	Dolomite
Nonmetallic luster	4	✔		colorless or variable	cleaves in 4 directions	hydrofluoric acid	CaF_2	Fluorite
Nonmetallic luster	5–6	✔		black to dark green	cleaves in 2 directions at 90°	mineral collections, jewelry	$(Ca,Na)(Mg,Fe,Al)$ $(Si,Al)_2O_6$	Pyroxene (commonly augite)
Nonmetallic luster	5.5	✔		black to dark green	cleaves at 56° and 124°	mineral collections, jewelry	$CaNa(Mg,Fe)_4(Al,Fe,Ti)_3$ $Si_6O_{22}(O,OH)_2$	Amphibole (commonly hornblende)
Nonmetallic luster	6	✔		white to pink	cleaves in 2 directions at 90°	ceramics, glass	$KAlSi_3O_8$	Potassium feldspar (commonly orthoclase)
Nonmetallic luster	6	✔		white to gray	cleaves in 2 directions, striations visible	ceramics, glass	$(Na,Ca)AlSi_3O_8$	Plagioclase feldspar
Nonmetallic luster	6.5		✔	green to gray or brown	commonly light green and granular	furnace bricks, jewelry	$(Fe,Mg)_2SiO_4$	Olivine
Nonmetallic luster	7		✔	colorless or variable	glassy luster, may form hexagonal crystals	glass, jewelry, electronics	SiO_2	Quartz
Nonmetallic luster	6.5–7.5		✔	dark red to green	often seen as red glassy grains in NYS metamorphic rocks	jewelry (NYS gem), abrasives	$Fe_3Al_2Si_3O_{12}$	Garnet

*Chemical symbols:
Al = aluminum, C = carbon, Ca = calcium, Cl = chlorine, F = fluorine, Fe = iron, H = hydrogen, K = potassium, Mg = magnesium, Na = sodium, O = oxygen, Pb = lead, S = sulfur, Si = silicon, Ti = titanium

✔ = dominant form of breakage

Reading the Table:

- **MINERAL NAME** – *far right*. This gives you the name of the mineral whose properties are being discussed.

- **COMPOSITION** – gives you the **chemical composition** of the mineral. The **key** to the symbols is on the *bottom* of the table.
  - **Ex:** A molecule of *hematite* (Fe_2O_3) is made up of two iron (Fe) atoms and three oxygen (O) atoms

- **LUSTER** – **shine** – *far left*. The minerals are divided into three groups:
  - **Metallic luster** – the mineral shines like a metal – includes all the minerals from *graphite* to *pyrite*
  - **Nonmetallic luster** – the mineral does not shine like a metal – includes all the minerals from *talc* to *garnet*
  - **Either** – the mineral sometimes shines like a metal and sometimes does not – only *hematite*

- **HARDNESS** – **resistance to being scratched**:
  - Measured on a scale from **1-10**; 1 – softest; 10 – hardest
  - Harder minerals can scratch softer minerals.

- **CLEAVAGE/FRACTURE** – tells you how the mineral usually breaks:
  - **Cleavage** – breaks evenly and smoothly
  - **Fracture** – breaks unevenly

- **COMMON COLORS** – gives you the common colors of this mineral. Notice that most minerals appear in more than 1 color. (Color cannot be used for identification because the mineral's natural color is often spoiled by impurities.)

- **DISTINGUISHING CHARACTERISTICS** – gives you identifying properties of the mineral.

- **USES** – gives you common uses of this mineral

> *Mineral* – an *inorganic naturally occurring solid* with definite chemical compositions. Minerals are the *building blocks* of most **rocks**.

Earth Science Reference Tables Page 16

118. The surface bedrock of New York State that is most likely to contain the mineral garnet can be found in an area 30 miles [See also RT 2-3 – Map of NYS]
    (1) north of Binghamton
    (2) south of Mt. Marcy
    (3) east of Oswego
    (4) west of Utica

119. Which rock is composed of a mineral that can be used for the production of cement? [See also RT 7.]
    (1) basalt
    (2) limestone
    (3) rock salt
    (4) rock gypsum

120. Which mineral is commonly mined as a source of the element lead (Pb)?
    (1) galena
    (2) quartz
    (3) magnetite
    (4) gypsum

Base your answers to questions 123 through 124 on the mineral chart below and on your knowledge of Earth science. The mineral chart lists some properties of five minerals that are the major sources of the same metallic element that is used by many industries.

Mineral Chart

Mineral Name	Composition	Density (g/cm³)	Hardness	Streak	Nonmetallic Luster	Common Colors
brucite	$Mg(OH)_2$	2.4	2.5-3	white	glassy to waxy	white
carnallite	$KMgCl_3 \cdot 6H_2O$	1.6	2.5	white	greasy	white
dolomite	$CaMg(CO_3)_2$	2.8	3.5-4	white	glassy to waxy	shades of pink
magnesite	$MgCO_3$	3.1	3.5-4.5	white	glassy	white
olivine	$(Fe,Mg)_2SiO_4$	3.3	6.5	white	glassy	green

121. Which two minerals have compositions that are most similar to calcite?
    (1) brucite and carnallite
    (2) carnallite and dolomite
    (3) dolomite and magnesite
    (4) magnesite and olivine

122. Which mineral might scratch the mineral fluorite, but would not scratch the mineral amphibole?
    (1) brucite
    (2) magnesite
    (3) carnallite
    (4) olivine

123. Which mineral has a different common color from its color in powdered form?
    (1) brucite
    (2) carnallite
    (3) magnesite
    (4) olivine

Earth Science Reference Tables Page 16

124. Base your answer to this question on the photographs below and on your knowledge of Earth science. The photographs show eight common rock-forming minerals.

Identify the two most abundant elements, by mass, in Earth's crust that are part of the composition of all eight of these minerals.
[See also RT 1 - Average Chemical Composition of Earth's Crust, Hydrosphere and Troposphere.]

_____   _____

125. Base your answer to this question on the cross section below and on your knowledge of Earth science. On the cross section, numbers 1 through 7 represent rock units in which overturning has not occurred. Line XY represents a fault and line WZ represents the location of an unconformity.

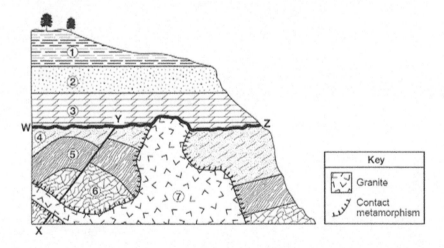

Write the chemical formula that shows the composition for the most common mineral found in rock unit 3. [See also RT 7.]

_____

ANSWERS

TO THE

ON THE

NEW YORK STATE

EARTH SCIENCE REFERENCE TABLES

Answers

Equations

1) 8° C/month (16/2 = 8)
2) 40 ft/mi (100/2.5 = 40)
3) any value from 1.0 in/h to 1.1 in/h (88/85 ≈ 1.04)
4) 15.44 g/15.4 g/15 g (19.3 X 0.8 = 15.44)

Specific Heats of Common Minerals

5) 4
6) 4

Properties of Water

7) 2
8) 3
9) 4,520 joules (2260 J/g X 2)

Average Chemical Composition of Earth's Crust, Hydrosphere & Troposphere

10) 1
11) 2
12) 4
13) 2

Generalized Landscape Regions & Geologic History of New York State

14) 2
15) 1
16) 1
17) 3
18) 1
19) 3 (Use latitude & longitude coordinates on RT 3, then move over to RT 2 and identify the region.)
20) 2 (Find Susquehanna on RT 3, then correspond map of NY State to given map.)
21) 4 (Correspond map to RT 2.)

Surface Ocean Currents

22) 1
23) 2
24) 1
25) California Current

Tectonic Plates

26) Nazca/Antarctic/Cocos/Caribbean Plate
27) convergent plate boundary/subduction zone/colliding plates
28) 4
29) 1
30) 3
31) 1
32) 1
33) convergent plate boundary/subduction zone/colliding

Answers

plates (See from diagram that two plates are colliding. Also see boundary between North American Plate near Alaska, far north, and the Pacific Plate.)

Relationship of Transported Particles to Water Velocity

34) 2
35) 4
36) 2
37) 2
38) 1.5 to 2.5 cm
39) *Water velocity:* increases/speeds up/gets greater/flows faster
    Amount of sediment: increases/becomes greater/less sediment is left behind in the sluice box

Scheme for Igneous Rock Identification

40) 1
41) 4
42) 3
43) 2
44) 4
45) 2
46)

Mineral Name	Felsic	Mafic
Potassium feldspar	X	
Olivine		X
Quartz	X	
Pyroxene		X

47) Plutonic, slow cooling rate
48) plagioclase feldspar or plagioclase/biotite or biotite mica/amphibole or hornblende/pyroxene or augite/quartz

Note: Quartz with either pyroxene or augite is not correct because a single andesite rock cannot contain both pyroxene and quartz.

Schemes for Sedimentary and Metamorphic Rock Identification

49) 2
50) 1
51) dolomite or $CaMg(CO_3)_2$
52) 2
53) 4
54) limestone or dolostone
55) 3
56) 3
57) 4

Inferred Properties of Earth's Interior

58) 3

Answers

59) 4
60) 4
61) 4
62) Continental crust: 2.7 g/cm³; oceanic crust: 3.0 g/cm³

Earthquake P-Wave & S-Wave Travel Time

63) 2
64) 3
65) 2
66) 2

Dewpoint/Relative Humidity

67) 1
68) 1 (difference is 2, wet bulb temperature is always less than or equal to dry bulb)
69) 3
70) *Dewpoint*: 8°C; *Relative humidity:* 40% (Dry bulb thermometer measures regular air temperature, so the "air temperature" is the dry bulb temperature.)

Temperature

71) 2

Pressure

72) 1

73) 3 (diagram is aneroid barometer – you must recognize it)
74) 30 in of Hg
75) *Pressure:* barometer; *temperature:* thermometer

Key to Weather Map Symbols

76) 3
77) 4
78) 3
79) 3
80) 4
81) 4
82) 2 (Air temperature and dewpoint are very similar → relative humidity is close to 100% → rain is likely.)
83) Note: 0.85 receives no credit because it is not in the proper weather station model format.
84) 3
85) The center of the X should be within the diagonally lined area shown on the map below:

Answers

86) any value from 39°F to 41°F (Watertown is in New York. First locate New York within this map of the USA. New York is on the east coast and is north, close to Canada – the area with no state divisions on the map. If you look carefully and compare this map to the map on RT 3, you'll notice that the body of water near the letter Z on the map is Lake Ontario. Watertown is on the east side of Lake Ontario – so its temperature must be close to Z's temperature – around the 40 range.)

87) Ocean; warmer OR hot OR a tropical temperature

88) Line should start from line AB, pass between the cold and warm labels, and curve up to the left, as shown in the diagram below.

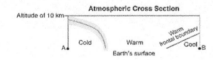

Selected Properties of Earth's Atmosphere

89) 1
90) 2
91) 3
92) Troposphere

Planetary Winds & Moisture Belts in the Troposphere

93) 3 (Answer is in the paragraph to the left of the Planetary Wind and Moisture Belts in the Troposphere table on RT 14.)
94) 4
95) 3
96) 1 (S.W. winds are coming from the southwest and moving towards the northeast.)
97) 4
98) 2
99) Barometer
100) NE/northeast/E/east
101) a straight or curved arrow drawn generally pointing toward the west or northwest as shown in diagrams below:

Electromagnetic Spectrum

102) 1
103) 1

Answers

Characteristics of Stars

104) 3
105) 2
106) 4
107) 2
108) 4

Solar System Data

109) 4
110) 3 (between Mars and Jupiter's periods of revolution)
111) 4
112) 2
113) 2
114) the Moon's period of rotation and period of revolution are equal OR the Moon rotates and revolves at the same rate/in the same amount of time OR the Moon rotates and revolves once in 27.3 days OR the Moon rotates only once per revolution
115) Mercury
116) Terrestrial planet; smaller diameter than Jovian OR higher density OR Terrestrial planet densities range from 3.9 g per cm³ to 5.5 g per cm³ OR rocky/solid OR not gaseous OR closer to the Sun OR less mass OR shorter period of revolution OR longer periods of rotation OR terrestrial planets don't have rings
117) the sun
118) 2 (First see mineral table on page 16 and see under the Distinguishing Characteristics column that garnet is often found in NYS metamorphic rocks. Then move to RT 2-3 and check out each of the choices, using the map scale on the bottom right of page 3 to determine the scale for 30 miles. 30 miles south of Mt. Marcy [near the Hudson River] is the only area composed of metamorphic rock.)

Properties of Common Minerals

119) 2
120) 1
121) 3
122) 2
123) 4
124) oxygen (O) and silicon (Si)
125) $CaMg(CO_3)$

MORE PRACTICE

- See the charts on the next two pages for additional practice regents questions, organized by table and by regents.
- Not all questions in the chart are based entirely on the reference tables. Some questions also require some background knowledge not included.
- Many questions may require the use of more than one table from the Earth Science Reference Tables. In that case, you'll find the question number listed twice, once under each table.
- By examining the table carefully, you'll notice that some tables appear on every regents multiple times (so they're worth reviewing!), while others have only appeared a few times in total on recent exams.
- To obtain the regents referenced on the chart, either use a recent Barron's review book or go to www.nysedregents.org/earthscience.

More Practice

PG		Jun 12	Aug 12	Jan 13	Jun 13	Aug 13	Jan 14	Jun 14	Aug 14
1	Equations	48	76	67	45		30, 70	56	62, 70
1	Specific Heats of Common Minerals					64			
1	Properties of Water						5		8
1	Average Chemical Composition of Earth's...	13	11	22				27	
2-3	Landscape/Geology of New York State	12, 16, 22, 56	18	32, 33	16, 60, 61	20	17	1, 2, 7, 24, 78	17, 22
4	Surface Ocean Currents	27, 84	38		9		10	51, 53	13, 81
5	Tectonic Plates	58, 66, 67, 85	16, 41, 61	19	6, 22, 71	18, 42, 43, 78, 79	18, 36	18, 20	21, 35, 83
6	Relationship of Transported Particles...		20	30	64, 65	23	61	77	58
6	Rock Cycle in Earth's Crust			50		26			
6	Scheme for Igneous Rock Identification	51, 52	22, 23, 24	20, 29, 49, 54, 56	66	17, 77, 82	81, 82	26, 28, 48	77, 78, 79, 8
7	Scheme for Sedimentary/Metamorphic Rock Identification	32, 53	31, 69, 70, 73	21, 29, 34, 49, 50, 54, 56	34, 48, 53	80	40, 41, 42, 57, 83	26, 78, 50	45, 53, 80
10	Inferred Properties of Earth's Interior	14	15	20	73	16, 17		19, 82, 83	20
11	Earthquake P-Wave & S-Wave Travel Time	31	14		21	19			
12	Dewpoint / Relative Humidity				31	12	8	58	12
13	Temperature				43	71			
13	Pressure								9
13	Key to Weather Map Symbols	23, 24, 68, 69, 75			8, 36, 38	31, 72, 83, 84, 85	6, 7, 11	41, 57	11, 31, 38, 3
14	Selected Properties of Earth's Atmosphere	83			13	9	37		57
14	Planetary Wind & Moisture Belts	27, 74	7			11, 15, 29		16, 53	14, 81
14	Electromagnetic Spectrum								3
15	Characteristics of Stars	3	3	36	7	5	44, 45	85	2
15	Solar System Data	2	27	84, 85	1	1	80	37, 38, 40	43, 72, 75
16	Properties of Common Minerals	20	21, 55	23	19, 66	34, 35		25, 27	23

More Practice

PG		Jan 15	Jun 15	Aug 15	Jan 16	Jun 16	Aug 16	Jan 17	Jun 17
1	Equations	76, 84	62, 71	76	70	77	82		64
1	Specific Heats of Common Minerals	21						12	
1	Properties of Water			8	65	43	4		12
1	Average Chemical Composition of Earth's...	31	10	34				84	
2-3	Landscape/Geology of New York State	9, 16, 68	13, 57	11, 13, 14, 18	5, 48	23, 45	35, 39, 40, 41	14, 78, 80, 81	29, 47
4	Surface Ocean Currents		65	50	17				21
5	Tectonic Plates	15, 62, 65	74	17, 61, 85	25, 53, 54, 56	35, 73, 74	59	22, 45	25, 60, 80
6	Relationship of Transported Particles...			7s5	75	79	80	49	28
6	Rock Cycle in Earth's Crust		31, 32						
6	Scheme for Igneous Rock Identification	64	45, 50	23	63	80, 81, 82	20, 22	47, 83	61, 79, 83
7	Scheme for Sedimentary/Metamorphic Rock Identification	24, 28, 55, 58	31, 33	21, 33, 63, 66, 67	31, 62, 78, 80	25, 50, 83	22, 25, 52	44, 79	
10	Inferred Properties of Earth's Interior	45	44	15		20	15	21	62
11	Earthquake P-Wave & S-Wave Travel Time		11	38	26	21	18	23	43
12	Dewpoint / Relative Humidity	12	6		14	11	64	10	10
13	Temperature								
13	Pressure					61			
13	Key to Weather Map Symbols	13, 17, 85	9, 52, 53	12		9, 10, 34	30	39	16, 70, 71
14	Selected Properties of Earth's Atmosphere	14	8	82	30		37	67	17
14	Planetary Wind & Moisture Belts	41				60	7		81
14	Electromagnetic Spectrum		37						1
15	Characteristics of Stars	80, 81, 82	41	1	42, 43	3	1	33	
15	Solar System Data	5, 6	1	55, 56, 57, 58	4, 27, 60		36, 75	5, 6, 77	51, 68
16	Properties of Common Minerals	22	34, 35		82, 83, 84, 85	52	42, 43, 44, 53	30, 82, 84, 85	29

Copyright 2019 © Y. Finkel | ALL RIGHTS RESERVED

Thanks for reading!

I'd love to hear how this book helped you understand any part of the Earth Science Reference Tables better than you did before.

Please email your comments or questions to unearthingnysesrt@gmail.com.

I welcome your feedback!

Y. Finkel

Made in United States
North Haven, CT
13 September 2023